4th International Conference on Materials Research and Innovation

IOP Conference Series: Materials Science and Engineering
Volume 1280

Bangkok, Thailand
15-16 December 2022

Editors:

Wirasak Smitthipong
Saree Phongphanphanee
Jirasak Wong-ekkabut

ISBN: 978-1-7138-7280-1
ISSN: 1757-8981

Printed from e-media with permission by:

Curran Associates, Inc.
57 Morehouse Lane
Red Hook, NY 12571

Additional copies of this publication are available from:

Curran Associates, Inc.
57 Morehouse Lane
Red Hook, NY 12571 USA
Phone: 845-758-0400
Fax: 845-758-2633
Email: curran@proceedings.com
Web: www.proceedings.com

TABLE OF CONTENTS

IOP Conf. Series: Materials Science and Engineering 1280 (2023) 011001 doi:10.1088/1757-899X/1280/1/011001

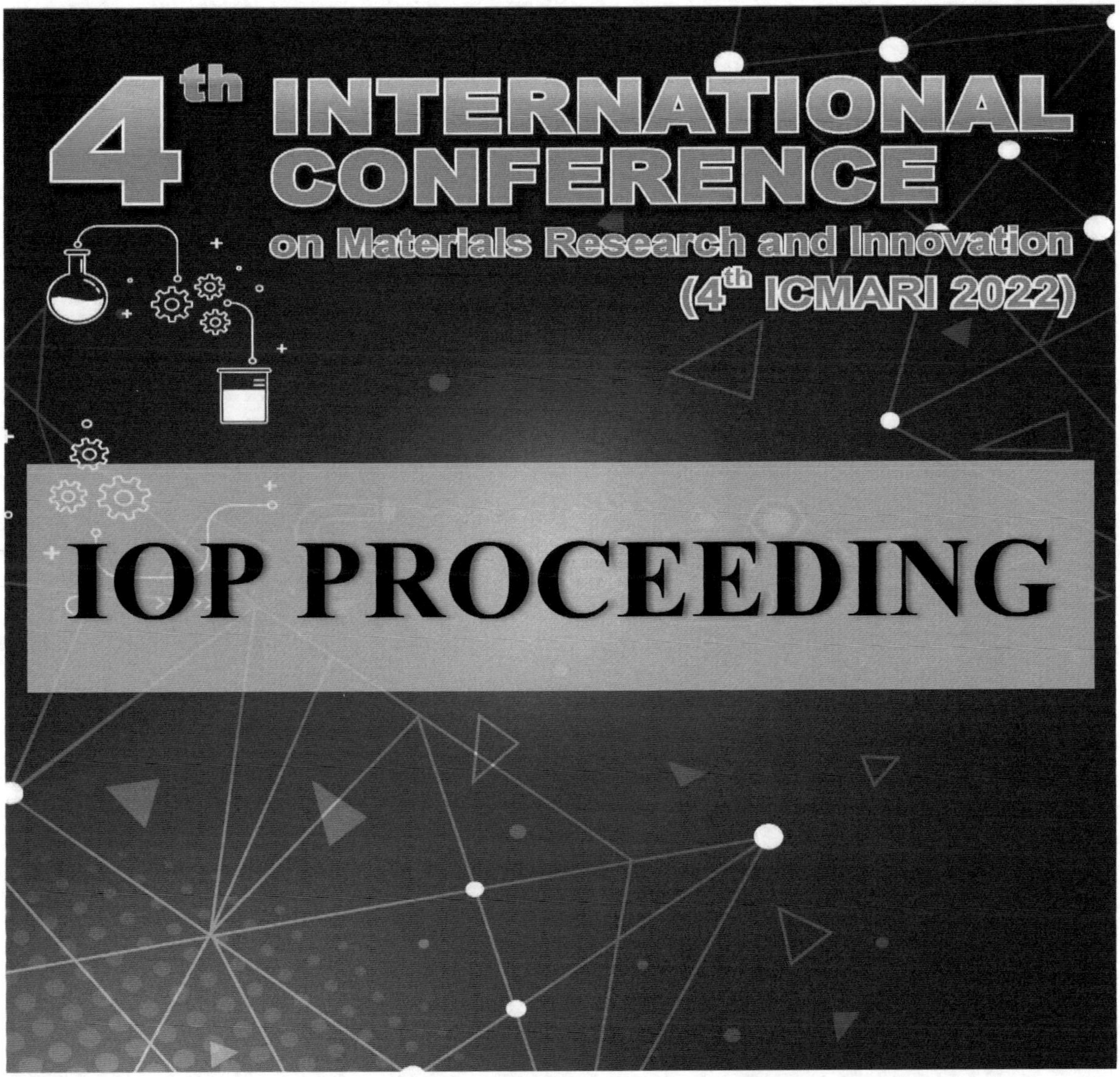

IOP Conf. Series: Materials Science and Engineering 1280 (2023) 011001 doi:10.1088/1757-899X/1280/1/011001

Preface

The 4th International Conference on Materials Research and Innovation (ICMARI 2022) was held on 15th – 16th December 2022, Bangkok, Thailand. The conference was conducted in frontier research on materials research and Innovation including Rubber and Polymeric Materials, Bio- and Circular-materials, and Special Advanced Materials. This 4th ICMARI was announced to focused on the Bio-Circular-Green Economic Model or BCG. The BCG has been applied by the research community and promoted by the Thai government as a new economic model for inclusive and sustainable growth, this BCG model also conforms with the UN Sustainable Development Goals (SDGs). The 4th ICMARI was a great opportunity for all participants to exchange knowledge and strengthen the research collaboration not only in ASEAN but also around the world. We hoped that the participants are fruitful by plenary and invited speakers from novel knowledge to new applications in the field of materials research and innovation.

Thank you all for participating,

ICMARI 2022 Organizing Committee

IOP Conf. Series: Materials Science and Engineering 1280 (2023) 011001 doi:10.1088/1757-899X/1280/1/011001

Organizing Committee

Assoc. Prof. Wirasak	Smitthipong	(Chairman)

Committee (Inside KU)

Asst. Prof. Saree	Phongphanphanee
Asst. Prof. Nattasamon	Petchsang
Asst. Prof. Decha	Dechtrirat
Assoc. Prof. Pongthep	Prajongtat
Assoc. Prof. Sutee	Boonchui
Assoc. Prof. Jirasak	Wong-ekkabut
Asst. Prof. Chomdao	Sinthuvanich
Asst. Prof. Wirunya	Keawwattana
Assoc. Prof. Kiadtisak	Saenboonruang
Assoc. Prof. Nanthiya	Hansupalak
Assoc. Prof. Prakit	Sukyai
Asst. Prof. Parames	Kamhangritirong
Dr. Supitta	Suethao
Dr. Hassarutai	Yangthong
Ms. Siriwan	Nerdnoi

Committee (outside KU)

Mr. Payab	Namprasert
Mrs. Pannee	Jirabunpaisit

Secretary team

Ms. Samaschaya	Namting
Mr. Kwanchai	Buaksuntear
Ms. Siriwan	Jansinak
Ms. Phakamat	Lim-arun
Ms. Piyanut	Promkaowthong
Ms. Haruthai	Klommueang
Ms. Weena	Anantawut
Ms. Ariya	Julbust

Editors of Proceeding

Assoc. Prof. Wirasak	Smitthipong	Chair Editor
Asst. Prof. Saree	Phongphanphanee	Editor
Assoc. Prof. Jirasak	Wong-ekkabut	Editor

IOP Conf. Series: Materials Science and Engineering 1280 (2023) 011001 doi:10.1088/1757-899X/1280/1/011001

International Technical Committee

Prof. Sadhan C.	Jana	University of Akron	USA
Assoc. Prof. Karine	Mougin	Institut de Science des Materiaux de Mulhouse	France
Prof. Arkadiusz	Chworos	Polish Academy of Sciences	Poland
Assoc. Prof. Wirasak	Smitthipong	Kasetsart University	Thailand
Asst. Prof. Parames	Kamhangritirong	Kasetsart University	Thailand
Asst. Prof. Saree	Phongphanphanee	Kasetsart University	Thailand
Asst. Prof. Nattasamon	Petchsang	Kasetsart University	Thailand
Prof. Hongbo	Zeng	University of Alberta	Canada
Prof. Atsushi	Takahara	Kyushu University	Japan
Prof. Katsura	Nishiyama	Meijo University	Japan
Assoc. Prof. Kheng-Lim	Goh	Newcastle University in Singapore	Singapore

IOP Conf. Series: Materials Science and Engineering 1280 (2023) 011001 doi:10.1088/1757-899X/1280/1/011001

Peer review statement

All papers published in this volume of *IOP Conference Series: Materials Science and Engineering* have been peer reviewed through processes administered by the proceedings Editors. Reviews were conducted by expert referees to the professional and scientific standards expected of a proceedings journal published by IOP Publishing.

IOP Conf. Series: Materials Science and Engineering 1280 (2023) 011002 doi:10.1088/1757-899X/1280/1/011002

Peer Review Statement

All papers published in this volume have been reviewed through processes administered by the Editors. Reviews were conducted by expert referees to the professional and scientific standards expected of a proceedings journal published by IOP Publishing.

- **Type of peer review:** Double Anonymous
- **Conference submission management system:** Morressier
- **Number of submissions received:** 30
- **Number of submissions sent for review:** 29
- **Number of submissions accepted:** 29
- **Acceptance Rate (Submissions Accepted / Submissions Received × 100):** 96.7
- **Average number of reviews per paper:** 2
- **Total number of reviewers involved:** 12
- **Contact person for queries:**
  Name: Jirasak Wong-ekkabut
  Email: jirasak.w@ku.th
  Affiliation: Kasetsart University

Antibacterial coating of corona treated Poly(lactic acid) / Poly(butylene succinate) film with Chitosan and Zinc oxide nanoparticle

N Hongsriphan*, J Nualyung, N Yaothaisong and P Patanathabutr

Department of Materials Science and Engineering, Faculty of Engineering and Industrial Technology, Silpakorn University, Sanam Chan Palace Campus, Nakhon Pathom 73000, Thailand.

*Corresponding author: Hongsriphan_N@su.ac.th

Abstract. The biodegradable PLA/PBS 90/10 wt% films were extruded and surface treated with corona discharge. Two electrical inputs of the corona treatment were used, 2 or 6 amperes, which the change of the surface tension was evaluated using corona measuring pens. The blend films were coated with the antibacterial mixture between chitosan and zinc oxide nanoparticles (ZnO-NP) in four concentrations: 1.0/0.4, 1.0/0.5, 1.5/0.5, and 1.5/0.5 wt%/wt%. This film could be used to replace transparent disposable packaging in the circular economy. It was found that the surface tension of the blend film was 30 dyne (mN/m) which was much like the neat PLA due to PLA being the main composition. Since the overall coated film thickness was kept constant, there was no significant difference in the coating weight among four coating concentrations. Compared to the uncoated PLA/PBS film, the tensile strength of the coated films tended to be higher with the increase of ZnO-NP content. However, the elongation at break of the coated films decreased significantly, which the stiffer coating layers became the stress concentrators. From SEM, it was seen that the good adhesion between the polymer and the coating layer was achieved. When the coated films were torn apart in the trouser tear testing, the failure was the breakage of the blend film with there was some coating layer delaminated.

1. Introduction

Poly(lactic acid) or PLA is one of the most important biodegradable plastics that replaces petroleum-based plastics for transparent disposable packaging in the circular economy [1]. However, its brittleness has to be improved by blending with other flexible biodegradable polymers or adding plasticizers. From our previous research [2], PLA was melt-blended with poly(butylene succinate) (PBS) in a weight ratio of PLA 90 wt% and PBS 10 wt% that was a transparent film with optimized mechanical properties. The film surface was corona treated to improve adhesion before coating with chitosan. Chitosan was selected to be a coating material due to its antibacterial activity and biobased material [3]. Another mineral that could be used to prevent the bacterial growth is zinc oxide nanoparticles (ZnO-NP). There have been many research articles [4, 5] that added chitosan or ZnO-NP into polymers that produced composites with antibacterial activity. In this research, the coating material was a mixture between chitosan and ZnO-NP in various ratios. The mixture was coated on the PLA/PBS films using a doctor blade coater, and evaluated mechanical properties. Prior to coating, the blend films were corona treated using two

IOP Conf. Series: Materials Science and Engineering 1280 (2023) 012001 doi:10.1088/1757-899X/1280/1/012001

different electrical input, and measured the surface tension by corona measuring pens. The coated films were also investigated the quality of the adhesion via the scanning electron microscope (SEM).

2. Experimental

2.1 Materials

Poly(lactic acid) or PLA (Ingeo™ Biopolymer 4032D, MFI of 7 g/10 min (210°C/2.16 kg)) was purchased from Nature Works LLC, USA. Poly(butylene succinate) or PBS (BioPBS™ FZ71PM, MFI of 22 g/10 min (190°C/2.16 kg)) was purchased from PTT MCC Biochem, Thailand. Low molecular weight chitosan with degree of deacetylation of 75-85% (no.448869) was purchased from Sigma-Aldrich. Zinc oxide nanoparticles (ZnO-NP, 99.5 % purity, molecular weight of 81.39 g/mole, average particles size of 20-40 nm) was kindly provided by Nano Materials Technology Co. Ltd, Thailand. Acetic acid (CH_3COOH, 99.8% purity) was purchased from Qchemical Co, Ltd, Thailand.

2.2 Preparation of Chitosan/ZnO coating material

Chitosan and ZnO-NP powder were dried and manually grinded. Two concentrations of the chitosan solution were prepared: 1.0 and 1.5 w/v%, which dried chitosan powder was added into a 1 w/v% acetic acid solution with continuous stirring at ambient temperature for 24 h. The obtained chitosan solution was filtered using a Buchner funnel/filter paper no.1. After that, the dried ZnO-NP powder of 0.4 or 0.5 wt% was added into the filtered chitosan solution and stirring continuously with a magnetic stirrer.

2.3 Cast film fabrication and corona treatment

PLA and PBS pellets were dried in an air-circulating over at 50°C for 12 h. The PLA/PBS blend was prepared with a fixed weight ratio of PLA 90 wt% and PBS 10 wt% using a co-rotating twin-screw extruder (SHJ-25, Nanjing Youngteng Chemical Equipment, China). The barrel/die temperature was 140-170°C and a screw speed was 80 rpm. The extrudate was cooled in water bath at ambient temperature and then pelletized. The cast film was fabricated using a cast film extruder (LBCR-150, Labtech Engineering Company Ltd., Thailand). The barrel/die temperature profile were 160-180°C, and the screw speed was 50 rpm. The chill-roll velocity was 2.9-3.0 m/min, and the nip-roll velocity was 3.4 m/min. The final thickness of the cast film was about 150 ± 50 µm. The PLA/PBS films were corona treated with a corona treater as shown in Figure 1A. The corona electrical input of the corona treatment was varied as 2 or 6 amperes, while the pull-out velocity was kept constant. The surface tension of the corona treated films were measured by means of corona measuring pens (DYNE TEST) as seen in Figure 1B. The number of the corona measuring pens were 30, 38, 42, 44, 46, 48, 50, and 52 dynes.

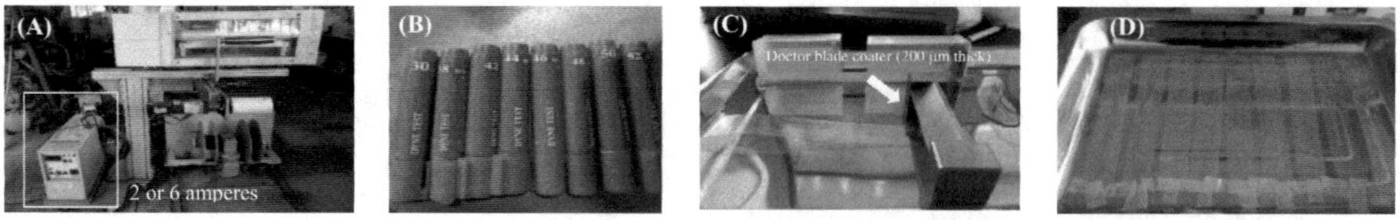

Figure 1. The corona treater (A), the corona measuring pen (B), the doctor blade coating (C), and the air drying of the coated films (D).

2.4 Film coating by a doctor blade

The film coating was performed using a doctor blade (Astro Instrument Ltd., Thailand) as shown in Figure 1C. The transparent PLA/PBS film was cut into film strips of 2.54 cm wide and 17.78 cm long to facilitate the coating process. Prior to coating, the film thickness and the film weight were measured individually. The corona treated films were placed under the doctor blade, and the coating solution was poured onto the film surface. The films were then pulled under the blade with a constant speed manually. The overall thickness of the wet coated PLA/PBS film was about 200 µm indicated by the gap between the doctor blade and the stainless-steel tray. The coated films were air dried at ambient temperature as shown in Figure 1D. The final dried film thickness was measured and calculated its coating thickness. They were also weighed and calculated the coating weight.

2.5 Tensile testing and Scanning electron microscopy (SEM)
The tensile testing of the uncoated and chitosan/ZnO coated films was carried out in accordance with ASTM-D882 using a universal testing machine (Instron 5969, Instron Engineering Corporation, USA). The specimens were cut by a cutter to obtain the rectangular strip of 2.54 x 15.24 cm. The gauge length was 10.16 cm. The crosshead speed was 5 mm/min. Seven trials were tested which the averages and standard deviations were calculated. Film specimens were soaked in liquid nitrogen, then were broken and examined using a scanning electron microscope (SEM) (Tabletop microscope TM3030, Hitachi, Japan). The surface was platinum-gold coated prior to inspection to avoid electrostatic charging.

3. Results and discussion
3.1 Surface tension of PLA/PBS film before and after corona treatment
The obtained blend film from PLA 90 wt% and PBS 10 wt% was transparent, which was an optimized blend composition from our previous research. From Figure 2, it was found that all film samples could be drawn a continuous line using a 30-dyne corona measuring pen, which the ink dried in 3 second and could not be wiped off. When the line was drawn using other corona measuring pens, a continuous line was not obtained after the ink dried in 3 seconds. This implied that all corona treated films in this research had similar surface tension characteristics because the main composition of the blend film was PLA. Since the corona measuring pen of 32 and 34 dyne were not available in the lab, the difference of the surface tension among the testing films could not be obtained.

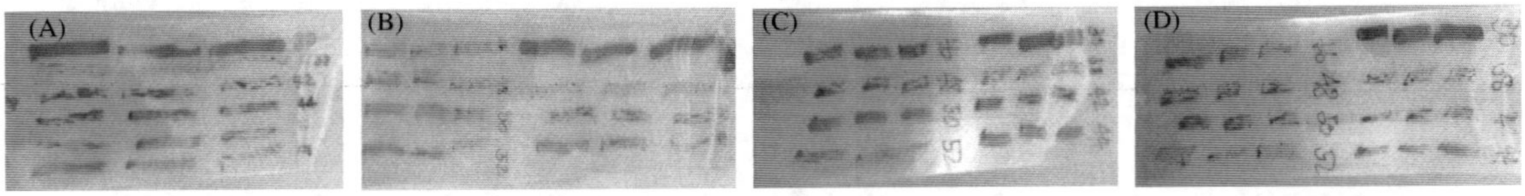

Figure 2. Surface tension measurement of corona treated PLA films using an electrical input of 2 ampere (A) and 6 ampere (B), and surface tension measurement of corona treated PBS/PLA films using an electrical input of 2 (C) and 6 amperes (D).

3.2 Film thickness and film weight
Table 1 presents film thickness and film weight of the corona treated PLA/PBS films coated with four chitosan/ZnO concentrations. It was found that the net coating thickness of all coating formula was about 0.01-0.02 mm. This was attributed to the uniform shrinkage of the coating layer after drying. Consider the coating weight, it was found that the film that was corona treated with an electricity input of 6 ampere had the highest net coating weight.

3.3 Tensile strength and elongation at break
Figure 3 presents tensile strength and elongation at break of corona treated PLA/PBS films coated with four chitosan/ZnO concentrations. The abbreviations of the film samples are coded as the following: 2A/C1.0/Z0.4 indicates the corona treated film with the electrical input of 2.0 ampere, and coated with chitosan of 1.0 wt% and ZnO-NP of 0.4 wt%. It was found that the elongation at break of the blend films reduced significantly after coating with chitosan and ZnO-NP. This was due to the constraints of motion by the rigid coating material. The tensile strength increased with the increase of the ZnO-NP content, and these particles became the stress concentrators to initiate the cracking of the blend films. The average of the coated films with the corona electrical input of 6 ampere was higher than those using 2 amperes. The trouser tear testing was performed in order to evaluate the film/coating layer adhesion quality. It was found that 67% of the film samples were broken in brittle behaviour resemble the breakage of PLA film, and 33% of the films were delaminated as shown in Figure 4. This implied the

good adhesion between the PLA/PBS film and the chitosan/ZnO coating. However, the tear strength of the untreated film was higher than the coated films. This confirms the dispersed ZnO particles on the coating layer became the starting points of crack propagation during the trouser tear testing. From SEM in Figure 6, it was seen that the good adhesion between the polymer and the coating layer was achieved. As our previous discussion [2] , the scheme of possible chemical interactions are the same.

Table 1. Film thickness and weight of corona treated PLA/PBS films with chitosan/ZnO coatings.

Corona electrical input (ampere)	Concentration		Film thickness (mm)		Net coating thickness	Film weight (g)		Net coating weight (g)
	Chitosan (wt%)	ZnO-NP (wt%)	Before coating	After coating		Before coating	After coating	
2.0	1.0	0.4	0.16 ± 0.02	0.18 ± 0.02	+0.02	0.7329 ± 0.0445	0.7384 ± 0.0445	+0.0055
2.0	1.0	0.5	0.15 ± 0.01	0.17 ± 0.01	+0.02	0.6407 ± 0.1031	0.6451 ± 0.1028	+0.0043
2.0	1.5	0.4	0.16 ± 0.02	0.18 ± 0.01	+0.02	0.7397 ± 0.0639	0.7489 ± 0.0639	+0.0092
2.0	1.5	0.5	0.13 ± 0.01	0.14 ± 0.01	+0.01	0.6905 ± 0.0698	0.6915 ± 0.0695	+0.0011
6.0	1.0	0.4	0.17 ± 0.02	0.18 ± 0.02	+0.01	0.7717 ± 0.1491	0.7888 ± 0.1330	+0.0171
6.0	1.0	0.5	0.15 ± 0.01	0.17 ± 0.02	+0.02	0.7074 ± 0.0516	0.7082 ± 0.0462	+0.0008
6.0	1.5	0.4	0.17 ± 0.01	0.18 ± 0.01	+0.01	0.7904 ± 0.0634	0.7955 ± 0.0629	+0.0051
6.0	1.5	0.5	0.16 ± 0.04	0.16 ± 0.01	~0.00	0.7167 ± 0.0739	0.7201 ± 0.0715	+0.0033

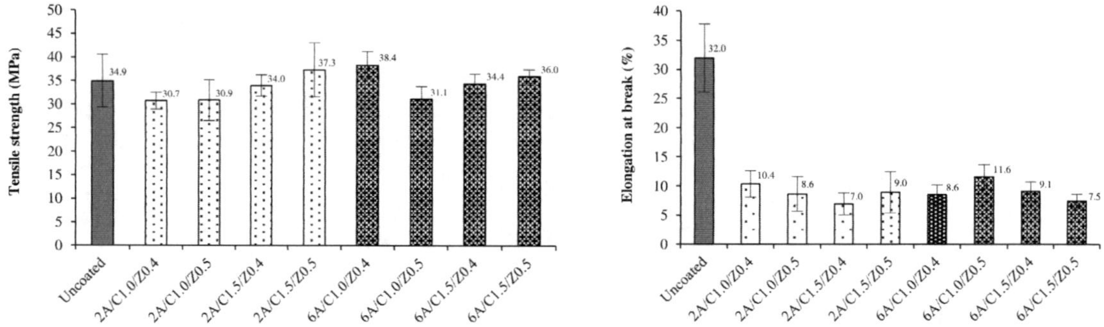

Figure 3. Tensile strength and elongation at break of corona treated PLA/PBS films with chitosan/ZnO coatings.

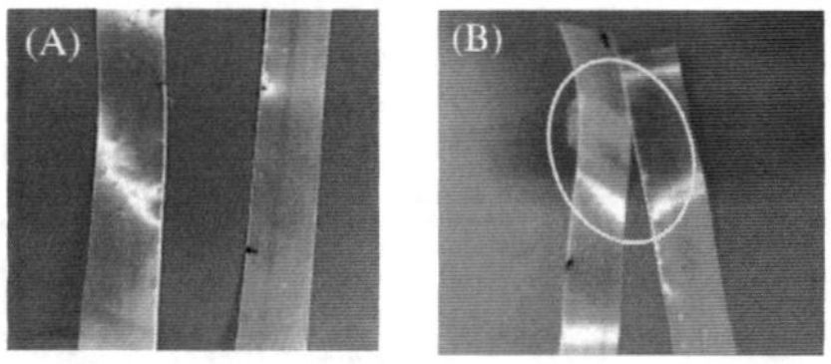

Figure 4. The brittle breakage of the coated film (A), the delamination of the coating layer from the coated film (B).

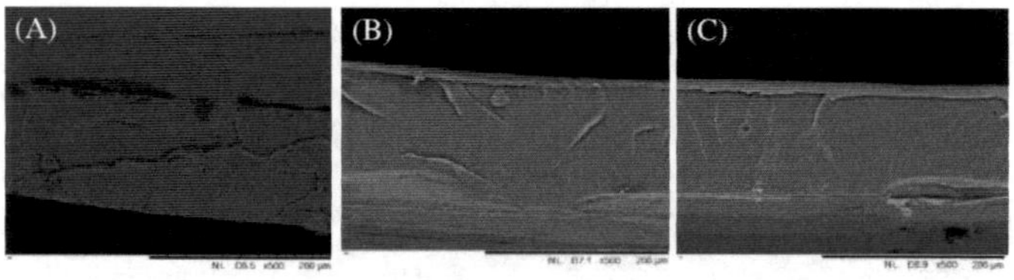

Figure 5. SEM micrographs (500X) of the uncoated PLA/PBS film (A), the PLA/PBS 2A/C1.5/Z0.5 (B), and the PLA/PBS 6A/C1.5/Z0.5 (C).

4. Conclusions

Using corona measuring pens, it was found that the surface tension of the biodegradable PLA/PBS 90/10 wt% films was 30 dyne (mN/m) which was much like the neat PLA due to PLA being the main composition. Since the coating thickness was kept constant, there was no significant difference in the comparison of the thickness and weight of the coating layers on the blend films. The tensile strength of the blend films was higher with the increase of the ZnO-NP content. When the coated films were performed the trouser tear testing, the failure was the breakage of the blend film with there was some coating layer delaminated. This implied good adhesion between the blend films and the mixture coating.

IOP Conf. Series: Materials Science and Engineering 1280 (2023) 012001 doi:10.1088/1757-899X/1280/1/012001

References

[1] Genovesi A, Aversa C, Barletta M, Cappiello G and Gisario A 2022 *Chem. Eng. Technol.* **6** 100419.

[2] Hongsriphan N and Sanga S 2018 *J. Plast. Film. Sheeting* **34** 160.

[3] Kumar S, Mukherjee A and Dutta J 2020 *Trends Food. Sci. Technol.* **97** 196.

[4] M.Youssef A and El-Sayed SM 2018 *Carbohydr Polym.* **193** 19.

[5] Al-Naamani L, Dobretsov S and Dutta J 2016 *Innov. Food Sci. Emerg. Technol.* **38** 237.

IOP Conf. Series: Materials Science and Engineering 1280 (2023) 012002 doi:10.1088/1757-899X/1280/1/012002

Application of the reference interaction site model self-consistent field method based on the Dirac–Hartree–Fock wave function to a chemical reaction

K Kanemaru[1], Y Watanabe[1], N Yoshida[2] and H Nakano[1,*]

[1] Department of Chemistry, Graduate School of Science, Kyushu University, 744 Motooka, Nishi-ku, Fukuoka 819-0395, Japan
[2] Department of Complex Systems Science, Graduate School of Informatics, Nagoya University, Furo-cho, Chikusa-ku, Nagoya 464-8601, Japan

[*]Corresponding author: nakano@chem.kyushu-univ.jp

Abstract. The reference interaction site model self-consistent field (RISM-SCF) method is a combined method of the electronic structure theory of molecules and the integral equation theory of molecular liquids. The RISM-SCF method based on the Dirac–Hartree–Fock wave function, recently proposed, is applied to a chemical reaction, specifically, a Menshutkin reaction in aqueous solution. The Helmholtz energy profile along the reaction coordinate is calculated and the characteristics of the reaction are discussed based on energy component analysis.

1. Introduction
Relativistic effects play an important role in the electronic structure of molecules containing heavy elements. It is now well known that both the scalar and spin–orbit effects are important for the geometry and properties of molecules. There are several methods to treat these relativistic effects theoretically. Among them, the four-component method [1] based on the Dirac equation is one of the most suitable methods for treating relativistic effects because it is based on the basic equation of relativistic quantum mechanics. Therefore, many methods for describing the electronic structure, including electron correlation methods, have been transferred to the relativistic four-component level.

For molecules in the solution phase, solvent effects also play an important role. Among the various solvation models, the polarizable continuum model (PCM) [2,3] is one of the most widely used approaches. In the PCM approach, the solvent molecules are considered as a continuum medium. In 2015, Di Remigio et al. [4] presented a formulation of a four-component relativistic self-consistent field (SCF) theory for a molecular solute described within the PCM for solvation, and successfully applied it to systems containing heavy elements. The PCM method is very effective, yet it does not include microscopic solute–solvent interactions such as hydrogen bonds. A method that can handle such microscopic interactions is the reference interaction site model self-consistent field (RISM-SCF) method [5,6], which combines the electronic structure theory with the reference interaction site model [7,8], an integral equation theory of molecular liquids.

IOP Conf. Series: Materials Science and Engineering 1280 (2023) 012002 doi:10.1088/1757-899X/1280/1/012002

Recently, we have proposed the RISM-SCF based on the Dirac–Hartree–Fock (DHF) method, one of the four-component relativistic molecular orbital (MO) methods [9]. This method can simultaneously calculate the precise electronic structure of molecules, including relativistic effects, with the DHF method and the solvent structure with the RISM theory. In the original article, the theoretical structure of the DHF/RISM-SCF method and its energy derivative is presented and applied to several molecular systems. In the present article, we present the application of this method to a chemical reaction involving heavy atoms—a Menshutkin reaction in aqueous solution, $NH_3 + CH_3I \rightarrow NH_3CH_3^+ + I^-$.

2. Method

2.1. DHF/RISM-SCF method

The DHF/RISM-SCF method [9] was obtained as a variational problem for free energy, as initially proposed by Sato et al. [10] for a multiconfiguration SCF wave function. Consider a system in which a quantum mechanical solute molecule is immersed in a solvent composed of classical molecules at infinite dilution. In the variational formulation, the basic equations are derived from the stationary conditions for the Helmholtz energy Lagrangian:

$$L\left(\mathbf{C}, \boldsymbol{\zeta}^{\mathrm{U}}, c, h, t\right) = E^{\mathrm{U}}(\mathbf{C}) + \Delta\mu(\mathbf{C}, c, h, t) + \boldsymbol{\zeta}^{\mathrm{U}} \cdot \boldsymbol{e}^{\mathrm{U}}(\mathbf{C}). \tag{1}$$

The first term of the right-hand side in Eq. (1) represents the electronic energy of the solute molecule and $\mathbf{C}$ is the set of variational parameters in the wave function. The second term represents the excess chemical potential of solvation, and c, h, and t are the direct, total, and indirect correlation functions describing the solvent structure around the solute molecule, respectively. The third term corresponds to constraints on the parameters $\mathbf{C}$ in the solute wave function, i.e., the orthonormality of orbitals and/or the normalization of the wave function. The symbols $\mathbf{e}^{\mathrm{U}}$ and $\boldsymbol{\zeta}^{\mathrm{U}}$ are the sets of constraints and multipliers, respectively. Taking variations of the Lagrangian with respect to the correlation functions and MO coefficients, we obtain the basic equations for RISM-SCF from the stationary conditions.

The resulting RISM-SCF equation is given as

$$\mathbf{F}^{\mathrm{solv}}\mathbf{C} = \mathbf{SC}\boldsymbol{\epsilon}, \tag{2}$$

where $\mathbf{F}^{\mathrm{solv}}$ is the four-component solvation Fock matrix

$$\mathbf{F}^{\mathrm{solv}} = \begin{bmatrix} \mathbf{F}^{\mathrm{LL,gas}} + \mathbf{V}^{\mathrm{LL,solv}} & \mathbf{F}^{\mathrm{LS,gas}} \\ \mathbf{F}^{\mathrm{SL,gas}} & \mathbf{F}^{\mathrm{SS,gas}} + \mathbf{V}^{\mathrm{SS,solv}} \end{bmatrix}, \tag{3}$$

and $\mathbf{C}$, $\mathbf{S}$, and $\boldsymbol{\epsilon}$ are the four-component coefficient, overlap, and eigenvalue matrices, respectively. The matrices $\mathbf{F}^{\mathrm{gas}}$ and $\mathbf{V}^{\mathrm{solv}}$ in Eq. (3) are the conventional Fock matrix in the DHF method and the potential matrix due to the solvent. By solving Eq. (2) coupled with the RISM equations, the electronic structure of the solute molecule and the solvation structure of the solvent are obtained simultaneously.

2.2. Computational details

The DHF/RISM-SCF method was applied to a Menshutkin reaction, $NH_3 + CH_3I \rightarrow NH_3CH_3^+ + I^-$, in aqueous solution. The Helmholtz energy profile was drawn along a reaction coordinate. The reaction coordinate was defined as the difference between the C–I and C–N distances. The basis sets used were the correlation-consistent polarized valence double-zeta (cc-pVDZ) basis sets for C, N, and H, and Dyall's double-zeta with polarization basis set for I. For CH_3I, the parameters for RISM were the same as given in our earlier paper [9]. For NH_3, the pairs of Lennard-Jones parameters (σ in Å, ε in kcal mol^{-1}) were (3.250, 0.1700) for N and (1.069, 0.0157) for H. The reaction path was determined by constrained optimization at the Hartree–Fock/cc-pVDZ-PP level.

3. Results and discussion

Figure 1(A) shows the Helmholtz energies A along the reaction coordinate $d = r(\text{C–N}) - r(\text{C–I})$ by the DHF/RISM-SCF and DHF/PCM-SCF methods, where A is shifted so that the energy of the reactant $NH_3 + CH_3I$ in the gas phase is zero. RISM-SCF gives a higher Helmholtz energy than PCM-SCF on

IOP Conf. Series: Materials Science and Engineering 1280 (2023) 012002 doi:10.1088/1757-899X/1280/1/012002

the reactant side while giving a lower Helmholtz energy on the product side. The reaction is exergonic, and the reaction free energies are −47.37 kcal mol^{-1} and −33.42 kcal mol^{-1} for RISM-SCF and PCM-SCF, respectively. This is because the solutes are ionized in the product state due to charge separation, which enhances the solute–solvent interaction and stabilizes the product state. RISM describes hydrogen bonds between solutes and solvents well, however, PCM underestimates them. RISM, therefore, has a greater stabilization of the product state compared with the PCM case. Because of the larger reaction free energy, the transition state (TS) of RISM-SCF is located more on the reactant side than in the case of PCM-SCF, at $d = 0.0$ Å for RISM-SCF and at 0.3 Å for PCM-SCF. The activation free energies are 7.23 and 12.94 kcal mol^{-1} for RISM-SCF and PCM-SCF, respectively. Note that an activation energy of 23.5 kcal mol^{-1}, determined by kinetic analysis, was reported for this reaction by Okamoto et al. [11,12]; it is different from the activation *free* energy, however, and therefore cannot be directly compared.

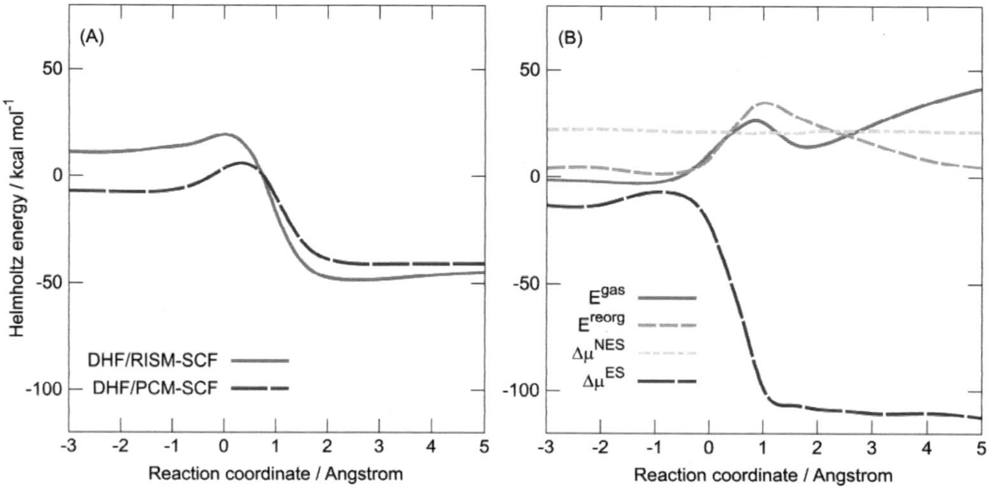

Figure 1. (A) Helmholtz energy profiles of RISM-SCF and PCM-SCF and (B) energy components of RISM-SCF.

The Helmholtz energy A can be decomposed into the solute energy in the gas phase E^{gas}, the reorganization energy E^{reorg}, and the nonelectrostatic and electrostatic parts $\Delta\mu^{\mathrm{NES}}$ and $\Delta\mu^{\mathrm{ES}}$ of the excess chemical potential, as $A = E^{\mathrm{gas}} + E^{\mathrm{reorg}} + \Delta\mu^{\mathrm{NES}} + \Delta\mu^{\mathrm{ES}}$. Figure 1(B) shows the components of A by RISM-SCF, where E^{gas} is shifted so that E^{gas} of the reactant is zero. E^{gas} has a very shallow first minimum and a second minimum, corresponding to the dipole compound and ion pair, respectively, as the typical Menshutkin reaction profiles in the gas phase show. The very shallow first minimum is due to the very small dipole moment of CH_3I. This energy is destabilized by the nearly constant $\Delta\mu^{\mathrm{NES}}$ and by E^{reorg}, although stabilized by $\Delta\mu^{\mathrm{ES}}$. The stabilization by $\Delta\mu^{\mathrm{ES}}$ is particularly pronounced after the TS, where the charges are separated.

Table 1 shows the Helmholtz energies A by RISM-SCF and their components at the reactant, TS ($d = 0.0$ Å), and the product. In the reactant, which is a pair of neutral molecules, E^{reorg} and $\Delta\mu^{\mathrm{ES}}$ are relatively small (3.88 and −13.55 kcal mol^{-1}, respectively) and $\Delta\mu^{\mathrm{NES}}$ is dominant (21.93 kcal mol^{-1}). In the TS, E^{gas} increases relative to the reactant (11.17 kcal mol^{-1}) and E^{reorg} also increases in response to the increase in charge bias (8.89 kcal mol^{-1}). $\Delta\mu^{\mathrm{ES}}$ also becomes large (−21.90 kcal mol^{-1}), as in E^{reorg}, but is still comparable to $\Delta\mu^{\mathrm{NES}}$ (21.33 kcal mol^{-1}) because the charges are not sufficiently well separated. This may be because the TS is on the reactant side in the Helmholtz energy profile. In contrast, in the product, E^{gas} increases significantly (95.27 kcal mol^{-1}) due to charge separation, whereas E^{reorg} is relatively small (0.87 kcal mol^{-1}) because the system has already been separated into ions. In the product, which is a pair of ions, $\Delta\mu^{\mathrm{ES}}$ is very large (–152.50 kcal mol^{-1}) and greatly stabilizes the system, outweighing the destabilization of E^{gas}.

Table 1 Helmholtz energies and their components at the reactant, TS, and product (kcal mol^{-1})

	Reactant		TS	Product	
	Total	$NH_3 + CH_3I$	NH_3–CH_3–I	Total	$NH_3CH_3^+ + I^-$
A [a]	12.26	–0.34 + 12.60	19.49	–35.11	
Component					
E^{gas} [a]	0.00		11.17	95.27	
E^{reorg}	3.88	3.15 + 0.73	8.89	0.87	0.87 + 0.00
$\Delta\mu^{\mathrm{NES}}$	21.93	7.46 + 14.47	21.33	21.25	13.95 + 7.30
$\Delta\mu^{\mathrm{ES}}$	–13.55	–10.95 – 2.60	–21.90	–152.50	–57.60 – 94.89

[a] The values of A and E^{gas} are shifted so that the energy of the reactant in the gas phase $E^{\mathrm{gas}}(\mathrm{R})$ is zero. $E^{\mathrm{gas}}(\mathrm{R}) = -4525521.55$ kcal mol^{-1}

Overall, the data in the figure and the table well illustrate the characteristics of the Menshutkin reaction, which is endergonic in the gas phase due to the ion dissociation, resulting in a late TS, but exergonic in the aqueous phase due to the large stabilization of the ionic product by the solvent, resulting in a early TS.

4. Conclusions

We have presented an application of the RISM-SCF method, based on four-component relativistic electronic structure theory, to a chemical reaction. We created Helmholtz energy profiles of a Menshutkin reaction $NH_3 + CH_3I \rightarrow NH_3CH_3^+ + I^-$ and conducted a component analysis. The reaction profile based on method well described the characteristics of the Menshutkin reaction. Component analysis revealed the contributions of the effects of electron reorganization, and electrostatic and nonelectrostatic interactions. The DHF/RISM-SCF method has proven to be an effective method for analyzing chemical reactions of systems containing heavy atoms such as iodine. The current method does not include the electron correlation effect in molecules. However, the development of RISM-SCF based on four-component electron correlation methods is currently underway.

References

[1] Dyall K G, Faegri K 2007, *Introduction to Relativistic Quantum Chemistry* (New York: Oxford University Press)
[2] Tomasi J, Mennucci B and Cammi R 2005 *Chem. Rev.* **105** 2999.
[3] Cossi M, Rega N, Scalmani G and Barone V 2003 *J. Comput. Chem.* **24** 669.
[4] Di Remigio R, Bast R, Frediani L and Saue T 2015 *J. Phys. Chem. A* **119** 5061.
[5] Ten-no S, Hirata F and Kato S 1993 *Chem. Phys. Lett.* **214** 391.
[6] Ten-no S, Hirata F and Kato S 1994 *J. Chem. Phys.* **100** 7443.
[7] Chandler D, Andersen H C 1972 *J. Chem. Phys.* **57** 1930.
[8] Hirata F 2004, *Molecular Theory of Solvation* (Dordrecht: Kluwer Academic Publishers)
[9] Kanemaru K, Watanabe Y, Yoshida N and Nakano H 2022 *J. Comput. Chem.* **44** 5.
[10] Sato H, Hirata F and Kato S 1996 *J. Chem. Phys.* **105** 1546.
[11] Okamoto K, Fukui S and Shingu H 1967 *Bull. Chem. Soc. Jpn.* **40** 1920.
[12] Okamoto K, Fukui S, Nitta I and Shingu H 1967 *Bull. Chem. Soc. Jpn.* **40** 2354.

Development of Bio-based Polybenzoxazine Composite Post : Effect of Glass Fiber Content on Mechanical Properties and Fracture Behaviors

P Mora[1], C Nunwong[1], S Rimdusit[2], U Srisorachatr[3], S Jampa-ngern[4] and C Jubsilp[1,*]

[1] Department of Chemical Engineering, Faculty of Engineering, Srinakharinwirot University, Nakhonnayok 26120, Thailand
[2] Research Unit in Polymeric Materials for Medical Practice Devices, Department of Chemical Engineering, Faculty of Engineering, Chulalongkorn University, Bangkok 10330, Thailand
[3] Institute of Dentistry, Department of Medical Services, Nonthaburi 11000, Thailand
[4] Department of Biomedical Engineering, Faculty of Engineering, Srinakharinwirot University, Nakhonnayok 26120, Thailand

*Corresponding author: chanchira@g.swu.ac.th

Abstract. This work aims to develop a novel composite post based on bio-based polybenzoxazine reinforced with glass fiber. The bio-based polybenzoxazine (poly(V-fa)) derived from vanillin, furfurylamine and paraformaldehyde, was prepared as polymer matrix in this study. Mechanical property and thermal property influenced by glass fiber contents were evaluated. The numerical simulation was also used to study the response to external applied load of tooth restored with glass fiber-reinforced poly(V-fa) composite posts by finite element analysis. From the results, not only the flexural properties, i.e. flexural strength and flexural modulus, were substantially improved up to 460 MPa and 25.6 GPa, respectively, but thermal properties, i.e. glass transition temperature (T_g) and degradation temperature at 5wt% loss (T_{d5}), of the composites were also enhanced with the reinforcement of glass fiber contents. Furthermore, the restoration of the developed glass fiber-reinforced poly(V-fa) composite post in term of the maximum von Mises stress and deformation areas was predicted via ANSYS program. The experimental and numerical results revealed that the bio-based polybenzoxazine composite reinforced with glass fiber have highly potential to be used as composite posts.
Keywords: Bio-based polybenzoxazine; Glass fiber; Post; Composites; Simulation

1. Introduction

Dental restorations are needed to resolve esthetic problems in patients who have stain due to intrinsic or extrinsic factors such as, antibiotics and smoking habits etc. The restorations required an opaque reinforcing structure know as posts [1]. Recently, various non-metal posts continuously replaced traditional metal posts due to their disadvantages, i.e. the risk of corrosion, root fractures, loss of retention [2]. Fiber posts have become increasingly popular in the dentistry industry due to its advantages in terms of ease of manipulation, mechanical qualities, esthetics, and removability, as well as their ability to give predictable clinical performance in a number of ways [2b].

Fiber posts basically know as composite materials that typically consist of reinforcing fiber bind with polymer matrix. The potential reinforcing fibers widely applied for composite post such as carbon fiber, polyethylene fiber, especially glass fiber etc. Mora et al. (2022) have developed fiber post based on glass fiber reinforced polybenzoxazine [2a]. They suggested that the fiber content in composite materials affected on the mechanical properties of their developed posts. Furthermore, they also studied the complex mechanical response to external applied load of the glass fiber-reinforced polybenzoxazine composites by using finite element analysis.

Polybenzoxazine are well known as a thermoset polymer which is continuously utilized as polymer matrix in fiber reinforced composites exhibiting high thermal properties and superior mechanical properties [3]. Furthermore, benzoxazine monomers exhibit very low melt viscosities, facilitating formulation with a wide variety of fillers or fillers.

Generally, petroleum resources are used as the basic ingredients for the synthesis of benzoxazine resin. With the increasing energy crisis and environmental problems around the world. The utilization of bio-based raw materials in the preparation of bio-based benzoxazine resin has become a hotspot for research. Benzoxazine resins have been synthesized using bio-based materials such as vanillin, eugenol, cardonol, furfurylamin, stearylamine, and guaiacol [4].

In this work, bio-based polybenzoxazine composite posts reinforced with glass fiber were developed. The polybenzoxazine (poly(V-fa)) were synthesized by using bio-based raw materials, which were vanillin, furfurylamine and paraformaldehyde. The effect of glass fiber contents on mechanical and thermal properties was evaluated. The numerical simulation was also used to study the response to external applied load of tooth restored with glass fiber-reinforced poly(V-fa) composite posts by finite element analysis in this work.

2. Materials and Methods

2.1. Materials

Paraformaldehyde (AR grade) was purchased from Merck Co., Ltd. (Darmstadt, Germany). Vanillin (99%) and furfurylamine ($\geq$99%) were supplied by Sigma-Aldrich Pte. Ltd. (Singapore). Glass fiber plain fabrics were obtained from Thai Poly Add Ltd. Partnership (Bangkok, Thailand).

2.2. Fabrication of Benzoxazine Resin, Prepreg and Composite

Benzoxazine resin (V-fa) was synthesized by vanillin, paraformaldehyde and furfurylamine at a molar ratio of 1:2:1 via solventless technology [5]. A dark brown liquid resin was obtained by mixing the three reactants for 40 min at 110°C.
The molten V-fa resin was pre-impregnated onto plain glass fabrics at 80 °C. The composite laminates were gently preheated at 150-170 °C for 3 h and cured at 180 °C for 2 h in the compression molder with a pressure of 15 MPa.

2.3. Bending Test

Flexural properties of glass fiber reinforced poly(V-fa) composite were determined according to ASTM D790M-93 by using a Universal Testing Machine. The specimen size was $50\times25\times3$ mm^3. A three-point bending test was performed with a support span of 48 mm using a crosshead speed of 1.2 mm/min.

2.4. Dynamic Mechanical Analysis

Dynamic mechanical property of the specimen were determined by a dynamic mechanical analyzer (model DMA1, Mettler Toledo) with three-point bending mode. The strain is measured at an amplitude of 30 μm with a frequency of 1 Hz. The sample was performed in the range of 30-300 °C at heating rate of 2 °C/min. The size of samples is 10 mm × 50 mm × 3 mm.

2.5. *Thermogravimetric Analysis*
Mettler-thermogravimetric Toledo's analyzer (model TGA1 Module) was utilized to study thermal stability of samples. The specimen was analyzed in the range of 25-800 °C at heating rate of 20 °C/min under a nitrogen atmosphere with flow rate of 50 mL/min.

2.6. *Finite Element Analysis (FEA)*
The response to external applied load of tooth restored of the developed composite was evaluated numerically with a commercial ANSYS Workbench 2022 R1 software. The material properties of the endodontically treated tooth model were obtained from the standard ANSYS AUTODYN material library. The material properties of the composites are input from the experimental results.

3. Results and discussion
3.1. *Mechanical Property*
Influence of glass fiber contents on flexural property of poly(V-fa) composite was investigated as shown in Figure 1. The flexural strength value of the composites was increased from 323 to 460 MPa with the addition of glass fiber contents in ranging from 50 to 60 wt%, respectively. The enhancement in flexural strength was due to a good interfacial bonding between glass fabric and polymer matrix [6], while, that of poly(V-fa) composite at 80 wt% glass fiber were slightly decreased to 430 MPa. The slight decrease in the strength values might be owing to micro-void formation which occurred from the less amount of polymer matrix to wet up the fabric. Moreover, the flexural modulus values of glass fiber- reinforced poly(V-fa) composites observed to increase with increasing glass fiber contents having values ranging from 9.8-25.6 GPa. It is, therefore, evident that an addition of much greater rigidity of glass fiber, i.e. 70-85 GPa [7] into the poly(V-fa) composites attributed to significant enhancement in the stiffness of the obtained poly(V-fa) composites [8]. Therefore, required elastic modulus should close to that of dentine (18 - 40 GPa [9]) which is a crucial parameter for load transmission to apply the glass fiber-reinforced poly(V-fa) composite as a glass fiber post.

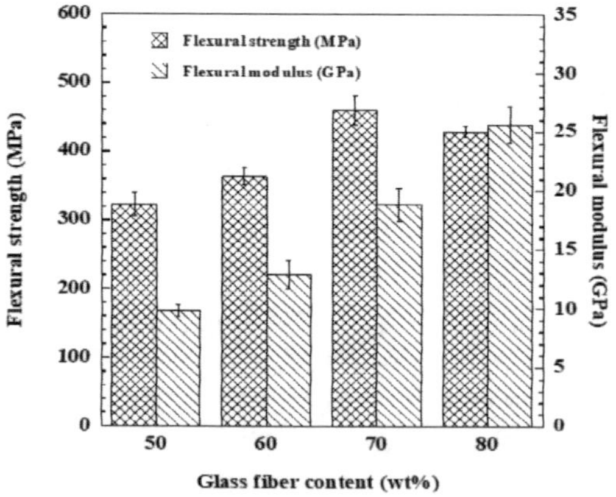

Figure 1. Flexural property of poly(V-fa) composites at various glass fiber content.

3.2. *Thermal Property*
Table 1 illustrates the dynamic mechanical property of poly(V-fa) composites reinforced with glass fiber contents ranging from 50 to 80 wt %. Storage modulus values at 30°C (the glassy state) of the composites increased with increasing glass fiber contents. These results are in a good agreement with the flexural properties mentioned above. Furthermore, thermal properties consisted of T_g and T_{d5} of of poly(V-fa) composites also tended to increase with increasing glass fiber loading. This increase in thermal properties of the composites might indicates the outstanding thermal properties of glass fabric as well as good interaction of the poly(V-fa) matrix to the glass fiber.

Table 1. Thermal and mechanical properties of poly(V-fa) composites at various glass fiber content.

Glass fiber content (wt%)	Storage modulus (GPa)	T_g (°C)	T_{d5} (°C)
50	12.5	164	442
60	15.4	167	429
70	18.0	170	464
80	19.8	170	594

3.3. Stress Analysis Using Finite Element Analysis

Figure 2 presents the von Mises stress distribution in a tooth restored with glass fiber reinforced poly(V-fa) composite post at 70 wt% glass fiber after load of 100 N with a occlusion at 45 degrees oblique is applied. From the results, the von Mises stress found at the center of the cervical region in dentine of the tooth restored having the composite posts similarly observed in an canal-treated tooth restored with fiber-reinforced composite, carbon fiber and glass fiber-reinforced PEEK by a finite element analysis [10].

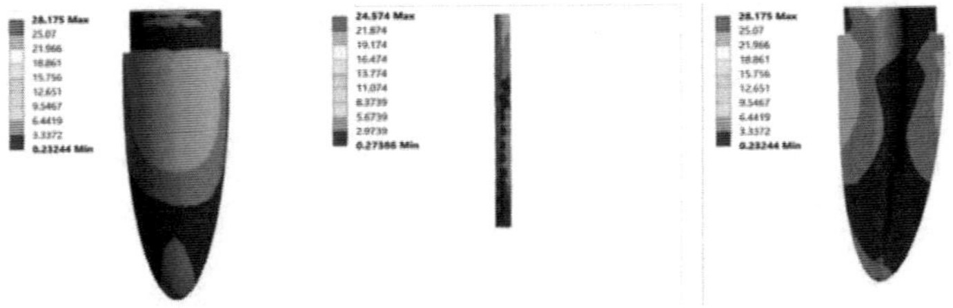

Figure 2. Deformation area of tooth restored of poly(V-fa) composites at 70 wt% glass fiber.

4. Conclusions

A novel composite post based on bio-based poly(V-fa) reinforced with glass fiber was developed in this work. The influence of glass fiber loading, i.e. 50, 60, 70 and 80 wt%, on mechanical and thermal properties was determined. From the results, flexural and thermal properties of the developed composites were substantially improved with the reinforcement of glass fiber. Furthermore, the numerical simulation revealed that the maximum von Mises stress and deformation areas occurred similar to that of the others. The experimental and numerical results revealed that the bio-based polybenzoxazine composite reinforced with glass fiber have highly potential to be used as composite posts.

Acknowledgement

This research has financial support from Thailand Science Research and Innovation 2022 (grant number 042/2565) and National Research Council of Thailand (NRCT) and Srinakharinwirot University (grant numbers N42A650377 and 588/2565)

References

[1] in *Chapter 4 - Fundamentals of Materials Science*, Eds.: R. Sakaguchi, J. Ferracane and J. Powers 2019 *Elsevier, Philadelphia*, pp. 29-68.

[2] a) Mora P, Nunwong C, Sriromreun P, Kaewsriprom P, Srisorrachatr U, Rimdusit S and Jubsilp C 2022 *Polymers* **14** 4321; b) Lamichhane A, Xu C and Zhang F Q 2014 *J. Adv. Prosthodont.* **6** 60.

[3] Ishida H 1996 in Process for preparation of benzoxazine compounds in solventless systems, Vol. US 5,543,516 1996.

[4] Prasomsin W, Parnklang T, Sapcharoenkun C, Tiptipakorn S and Rimdusit S 2019 *Nanomaterials* **9** 881.

[5] Ishida H 1996 in *Process for preparation of benzoxazine compounds in solventless systems*, *Vol.* (Ed. US Pat. 5, 516 (1996).
[6] Pol M H and Liaghat G 2016 *Polym. Compos.* **37** 1173.
[7] Jacobs M J N and Dingenen J L. J V 2001 *J. Mater. Sci.* **36** 3137.
[8] Wypych G 2016 in *8 - The Effect of Fillers on The Mechanical Properties of Filled Materials*, (Ed. G. Wypych), *ChemTec Publishing*, pp. 467-531.
[9] Ausiello P, Gloria A, Maietta S, Watts D C and Martorelli M 2020 *Polymers.* **12** 1836.
[10] Nahar R, Mishra S K and Chowdhary R 2020 *J. Oral. Biol. Craniofac. Res.***10** 719.

Optimization of microwave-assisted biodiesel production using Iron (III) oxide - Calcium oxide / Activated charcoal derived from waste Asian green mussel shell as heterogeneous catalyst

V Loryuenyong, S Kaewmanee, S Rattanawaraporn, N Chimplenapanon and A Buasri*

Department of Materials Science and Engineering, Faculty of Engineering and Industrial Technology, Silpakorn University, Nakhon Pathom 73000, Thailand.

*Corresponding author: achanai130@gmail.com

Abstract. The primary objective of this study was to produce a heterogeneous iron (III) oxide (Fe_2O_3)-calcium oxide (CaO)/activated charcoal (AC) catalyst from Asian green mussel shell utilizing wet impregnation and calcination processes for biodiesel synthesis. X-ray diffraction (XRD), scanning electron microscopy (SEM), energy dispersive X-ray spectroscopy (EDX), and the Brunauer-Emmett-Teller (BET) method were used to analyze the novel catalyst. Using a central composite design (CCD)-based response surface methodology (RSM), the process parameters such as catalyst quantity, reaction time, and microwave electrical power were optimized. The optimal reaction conditions were determined to be 1.07 wt.% of Fe_2O_3-CaO/AC, 6.42 min of reaction time, and 667 W of microwave electrical power. The published results imply that the heterogeneous catalyst has a great potential for the direct conversion of waste frying oil (WFO) to biodiesel, with the ability to reuse without reactivation. The development of Fe_2O_3-CaO/AC as a novel heterogeneous catalyst would facilitate the manufacture of biodiesel in an eco-friendly manner.

1. Introduction

There are several ongoing issues associated with the production, distribution, and retail pricing of non-renewable energy sources including fossil fuels and petro-diesel, as well as the emission of greenhouse gas (GHG). If we do nothing to address these problems, they will only become worse, leading to ever more severe energy shortages, climatic change, and global warming. As a result, researchers all around the world have been increasingly interested in these pressing challenges, which has stimulated the development of new forms of alternative energy. The most effective way to deal with this issue is to switch to biodiesel from conventional, polluting fuels like gasoline and diesel [1].

A variety of catalysts, including acid and alkaline catalysts, have been used in the past for biodiesel synthesis by transesterification. Several major drawbacks, such as environmental contamination and separation difficulties, have limited the adoption of conventional homogeneous catalysts. High reaction temperatures, long reaction times, and high reaction pressures have limited the growth and use of acid catalysts in industry in recent years [2]. For the most part, basic catalysts provide more conversion at lower temperatures than solid acids. As an active catalyst, calcium oxide (CaO) is commonly used in the transesterification process in these heterogeneous catalysts. Numerous efforts have been made to boost the output of biodiesel by employing various CaO catalysts. [3-4]. The goal of this work is to make the transesterification reaction of waste frying oil (WFO) more effective by adding a heterogeneous mixed metal iron (III) oxide-calcium oxide (Fe_2O_3-CaO)/activated charcoal (AC) catalyst to the reaction system. This catalyst could act as an acid-base catalyst and speed up both the esterification and transesterification processes at the same time.

2. Experimental

The Asian green mussel shells were collected as waste from a farm in Thailand's Samut Songkhram Province. Iron (III) nitrate nonahydrate and AC were purchased from A.C.S Xenon Limited Partnership (Thailand). The WFO feedstock was obtained at no cost from various restaurants in the neighbourhood of Silpakorn university in Nakhon Pathom, Thailand. It was filtered first to remove the contaminants, and then it was heated in the oven at a temperature of 100 °C for 30 min to remove any remaining moisture. All the remaining compounds were reagents of an analytical grade (Merck, greater than 99 % purity), and they were all utilized in their natural state.

CaO was made from waste Asian green mussel shells so that it could be used as a good and cheap source of calcium. First, the waste shells were washed with water, and then they were dried at 60 °C for 24 h. The shells were then put in an oven at 1,000 °C for 2 h until they were calcined. After that, they were ground and passed through a 60-140 mesh sieve (106-250 μm) to get powder. A wet impregnation method was utilized in order to produce the Fe_2O_3-CaO nanocomposite [5]. For the production of the Fe_2O_3-CaO/AC catalyst, metal oxides were combined with AC at a mass fraction ratio of 3:1 [6]. Before being used, each sample was stored in a closed container to avoid any potential interactions with the carbon dioxide (CO_2) and humidity that were present in the air. The novel catalyst was characterized by means of X-ray diffraction (XRD), scanning electron microscopy (SEM), energy dispersive X-ray spectroscopy (EDX), and the Brunauer-Emmett-Teller (BET) technique.

By means of central composite design (CCD) of response surface methodology (RSM), the effect of transesterification reaction parameters such as catalyst quantity (0.5-1.5 wt.%), reaction time (5-9 min), and microwave electrical power (500-700 W) on %WFO conversion was investigated. Each test was conducted 3 times, and the standard deviation for any given point never exceeded 7%. Using Design-Expert® Software, version 13 (Stat-Ease, Inc., Minneapolis, MN, USA), a quadratic polynomial equation was generated. The analysis of variance (ANOVA) test was utilized to assess the goodness-of-fit of the formulated mathematical equation model [7]. The biodiesel's composition was evaluated using gas chromatography-mass spectrometry (GC-MS).

3. Results and discussion

XRD pattern of the novel catalyst is shown in Figure 1. The presence of characteristic peaks at 2θ = 36.10°, 39.67°, 57.88°, 58.15° and 64.92° suggests that the Fe_2O_3 was effectively synthesized, in line with earlier findings [8]. The orientations (1 1 1) and (2 0 0) of the CaO phase were thought to be responsible for the highly noticeable and strong peaks of material that occurred at 2θ = 29.14° and 43.37°, respectively [9]. The peak at 2θ = 22.31° corresponds to the AC peaks. The SEM image of the catalyst revealed a good dispersion of Fe_2O_3-CaO on the AC surface. These results show that AC kept its structure, which is important for catalysis, and that the iron and calcium species were evenly spread across the surface of the support [10]. Figure 2 shows the EDX spectrum and the types of elements that make up the catalyst. With atomic percent values of 15.10, 2.55, 22.70, and 59.65, respectively, calcium, iron, carbon, and oxygen were clearly the main elements in the catalyst. The metal load would be deposited on the surface of the catalyst rather than the catalyst pore during the impregnation process, as seen in the SEM results. The pore volume of the Fe_2O_3-CaO/AC catalyst was 0.119 cm^3/g, and its BET surface area was 278.3 m^2/g. The considerable decrease in the BET surface area of the catalyst shows that mixed metal oxide molecules have filled the pores of the AC, supporting the concept. Moreover, it is obvious that the catalyst significantly increased the number of active sites [11].

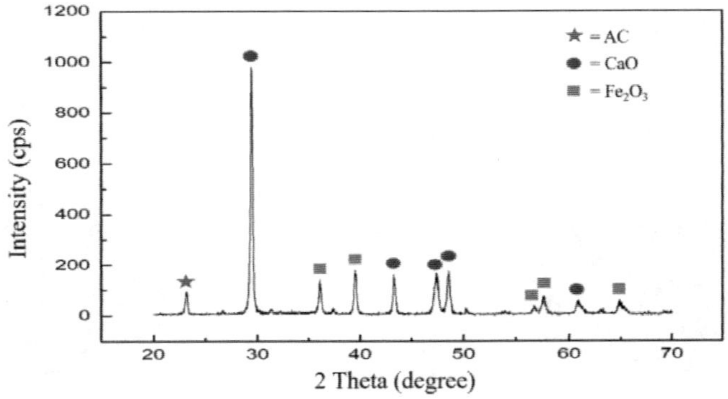

Figure 1. XRD pattern of the Fe_2O_3-CaO/AC catalyst.

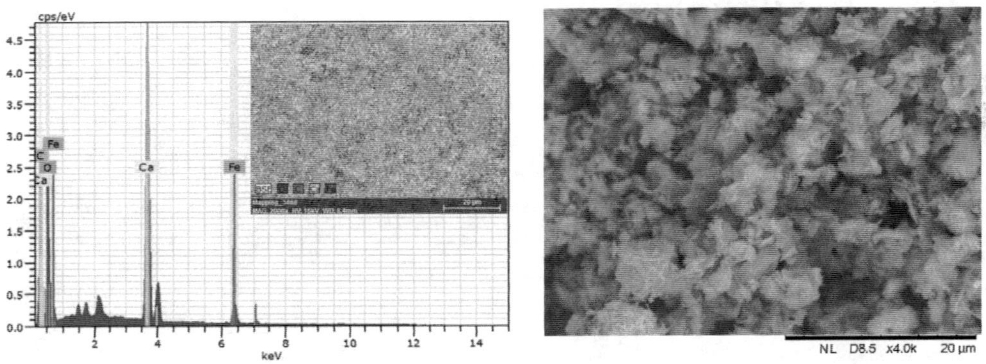

Figure 2. SEM image and EDX analysis of the Fe_2O_3-CaO/AC catalyst.

Transesterification experimental data was analyzed statistically using the Design-Expert® 13 software tool. It has been determined that the catalyst quantity (A), reaction time (B) and microwave power (C) are the most important input factors for this reaction. To illustrate the influence of the various reaction parameters on the %WFO conversion in a chemical reactor, we employ three-dimensional (3D) surface plots. Figure 3 displays the results of a regression model used to determine the ideal reaction conditions based on the limit criterion of optimizing the reaction conversion while minimizing the quantity of catalyst consumed [12-13], the amount of time spent on the reaction, and the amount of energy expended. The optimum parameters for %WFO conversion include a catalyst quantity of 1.07 wt.%, a reaction time of 6.42 min, and a microwave power of 667 W.

4. Conclusions

Finally, a new heterogeneous catalyst, was effectively synthesized from waste Asian green mussel shell and used to the conversion of WFO to biodiesel. Under optimal working circumstances, i.e. 1.07 wt.% catalyst amount in 6.42 min at 667 W microwave electrical power, the developed catalyst was able to catalyze the conversion of WFO by more than 98%, with no interruption from saponification or emulsification issue. Fe_2O_3-CaO/AC, a novel magnetic heterogeneous alkaline catalyst, shows tremendous promise as a component in biodiesel synthesis due to its acid-resistance and stability in repeated reuse.

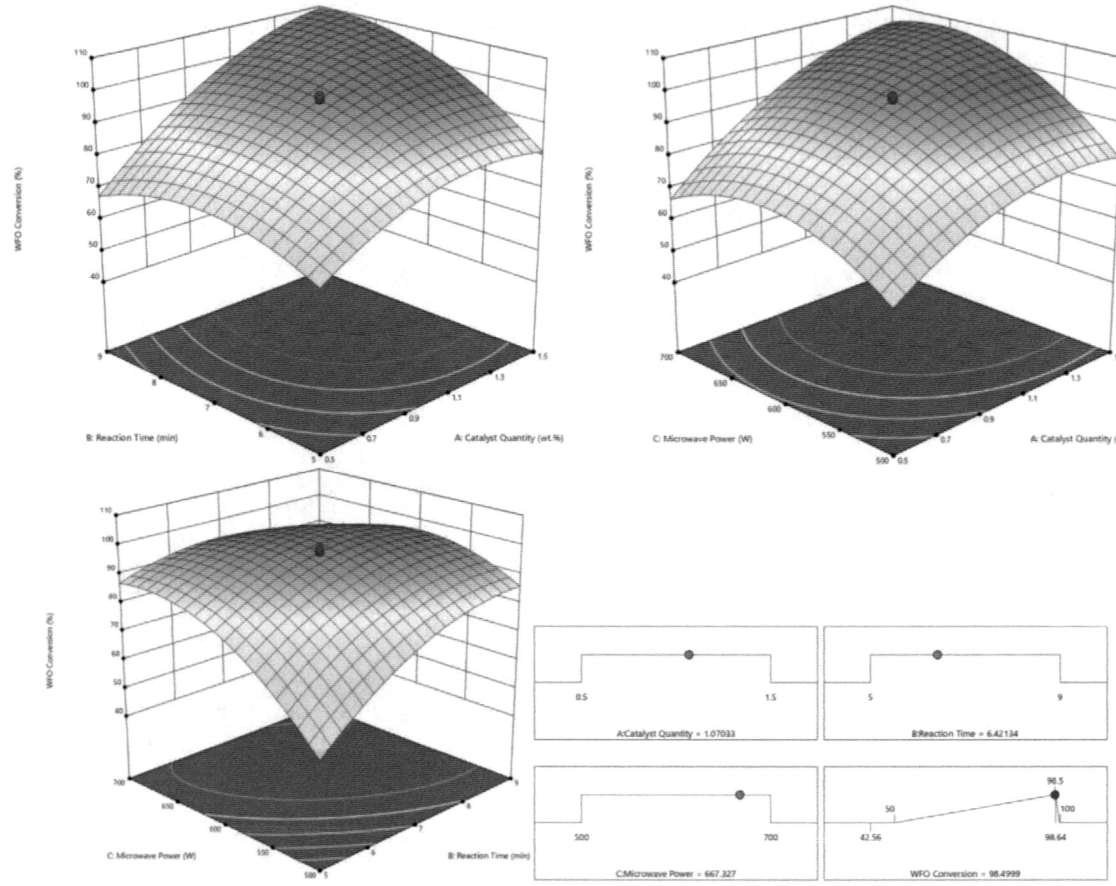

Figure 3. Optimal reaction conditions by RSM plots of the catalyst quantity, reaction time and microwave power effects on %WFO conversion.

References

[1] Zul NA, Ganesan S, Hamidon TS, Oh W-D and Hussin MH 2021 *J. Environ. Chem. Eng.* **9** 105741.
[2] Shi M, Zhang P, Fan M, Jiang P and Dong Y 2017 *Fuel* **197** 343.
[3] Liu Y, Zhang P, Fan M and Jiang P 2016 *Fuel* **164** 314.
[4] Ezzah-Mahmudah S, Lokman IM, Saiman MI and Taufiq-Yap YH 2016 *Energy Convers. Manag.* **126** 124.
[5] Keshavarz M, Foroutan R, Papari F, Bulgariu L and Esmaeili H 2021 *Sep. Sci. Technol.* **56** 1328.
[6] Buasri A, Chaiyut N and Nakweang C 2011 *Chiang Mai J. Sci.* **38** 572.
[7] Balajii M and Niju S 2019 *Energy Convers. Manag.* **189** 118.
[8] Helmiyati H and Masriah I 2019 *AIP Conf. Proc.* **2168** 020062.
[9] Buasri A, Lukkanasiri M, Nernrimnong R, Tonseeya S, Rochanakit K, Wongvitvichot W, Masa-ard U and Loryuenyong V 2016 *Korean J. Chem. Eng.* **33** 3388.
[10] Baroutian S, Aroua MK, Raman AAA and Sulaiman NMN 2010 *Fuel Process. Technol.* **91** 1378.
[11] Buasri A and Loryuenyong V 2017 *Mater. Today: Proc.* **4** 6051.
[12] Mostafaei M, Ghobadian B, Barzegar M and Banakar A 2015 *Ultrason. Sonochem.* **27** 54.
[13] Pham EC, Le TVT, Le KCT, Ly HHH, Vo BNT, Nguyen DV and Truong TN 2022 *Energy Rep.* **8** 5739.

Effects of Lignocellulose Contents in Bamboo Charcoal on Mechanical and Thermal Properties of Polybenzoxazine-based Friction Materials

N Jantapanya[1], C Jubsilp[2], P Mora[2], S Kaewluan[3], S Srisorrachat[4] and S Rimdusit[1,*]

[1] Research Unit in Polymeric Materials for Medical Practice Devices, Department of Chemical Engineering, Faculty of Engineering, Chulalongkorn University, Bangkok, 10330, Thailand.
[2] Department of Chemical Engineering, Faculty of Engineering, Srinakharinwirot University, Nakhonnayok 26120, Thailand.
[3] Department of Mechanical Engineering, Faculty of Engineering, Srinakharinwirot University, Nakhonnayok 26120, Thailand
[4] Department of Civil and Environmental Engineering, Faculty of Engineering, Srinakharinwirot University, Nakhonnayok 26120, Thailand

*Corresponding author: sarawut.r@chula.ac.th

Abstract. The objective of this research is to study mechanical and thermal properties of polybenzoxazine-based friction materials (PFMs) filled with 10 wt% bamboo charcoal particles (BC) containing residual lignocellulose content of 0-10wt% (BC0-10%), 10-20wt% (BC10-20%) and 20-30wt% (BC20-30%). The PFMs filled with synthetic graphite particles was also prepared to compare the properties. The obtained results showed that the mechanical properties, i.e., flexural strength and flexural modulus of the PFMs were improved with an increase of the residual lignocellulose content, and tended to be higher than that of the PFMs filled with synthetic graphite particles. Thermal stability monitored by the degradation temperature at 5% weight loss (T_{d5}) of the PFMs revealed a decrease with an increasing residual lignocellulose content in BC. However, the T_{d5} of the PFMs filled with BC is still higher than the generated heat during the braking, which has temperatures around 350 °C. The results suggested that the obtained composite material is a promising candidate for car brake pads.

1. Introduction

Due to increasing environmental concerns, the bio-circular-green economy model (BCG) has been gaining traction and increasing interest in the research field. The BCG model encompasses the use of renewable biological resources into value-added products. Many attempts have been made to utilize the bio-based materials as a replacement for synthetic materials to enhance the industry's long-term competitiveness. In this study, the focus is on using bamboo charcoal as an alternative friction modifier to substitute synthetic graphite in brake pad.

Brake pads are an essential component of vehicles. Through friction, brake pads convert kinetic energy into thermal energy. The friction materials must have a stable coefficient of friction and a low rate of wear across a wide range of operating speeds, pressures, temperatures, and environmental conditions, as well as high thermal stability and good mechanical properties. [1, 2]. Brake pads consist of binder resin, fiber reinforcement, solid lubricants, abrasives, and fillers [1]. Non-asbestos organic friction materials (NAO FMs) utilizing benzoxazine resins as polymeric binders have been increasingly developed due to good interfacial adhesion with fillers, high thermal stability, and excellent mechanical properties that result in benzoxazine resins suitable for use as polymeric binders for NAO FMs [3]. In recent years, agricultural waste such as bananas and rice husks has been used as reinforcing materials due to their high fiber content [4]. In addition, bamboo fiber was used as a material in motorcycle brake pads [5]. It was found that the composite has good mechanical properties with a hardness of 91.8 HRR and good heat resistance. The effect of the addition of lignocellulosic banana fiber on mechanical properties of automobile brake pads has been studied [6]. It was found that the lignocellulose-containing banana fibers can react with phenolic resin to give the material strength.

Therefore, in this research, the effects of residual lignocellulose contents in bamboo charcoal on mechanical and thermal properties of polybenzoxazine-based friction materials (PFMs) were studied. The residual lignocellulose mass content in bamboo charcoal in this research was in a range from 0-10 wt%, 10-20 wt%, and 20-30 wt%.

2. Experimental

2.1 Synthesis of benzoxazine resin (BA-35x)

The method for synthesizing benzoxazine resin (BA-35x) is by using a solventless technique [7]. The molar ratio of bisphenol A, paraformaldehyde, and 3,5-xylidine were 1:4:2. All reactant were mixed at 110 °C for 40 mins until a clear yellow liquid mixture was obtained and then left at room temperature to obtain a clear yellow solid. Finally, bring it to a fine powder and store it in the refrigerator.

2.2 Sample preparation

The formulation of the polybenzoxazine-based friction materials with different lubricants (BCx-y% with x-y% denoting the range of residual lignocellulose mass content in the bamboo charcoal and synthetic graphite) is shown in Table 1. All ingredients were mixed for 40 mins at 125 °C with an internal mixer at a rotor speed of 40 rpm. The compounds were then compressed for 2 h in a compression molder at 200 °C under 15 MPa of hydraulic pressure. Allow the sample to cool to room temperature and continue for analysis.

Table 1. Formulation of polybenzoxazine-based friction materials.

Formula	Binder (Benzoxazine resin) (wt%)	Reinforcing fibers (wt%)	Fillers (wt%)	Friction additives (wt%)	
				Abrasive	Lubricants
PFMs/Lubricants	10	40	33	7	10

2.3 Characterizations

Fourier transform infrared (FTIR) spectra of bamboo charcoal were obtained in transmittance mode (Perkin Elmer Spectrum GX) from 4000 to 400 cm^{-1} for a total of 64 scans. Thermal properties including degradation temperature at 5% weight loss (T_{d5}) and residual weight at 1000 °C of the samples were analyzed using a thermogravimetric analysis (TGA1, Mettler Toledo) under N_2 flow rate of 50 mL/min. The sample with mass of 15-20 mg was heated from 25 to 1000 °C with a heating rate of 20 °C/min. Mechanical properties of the samples were tested using a universal testing machine (model 5567 from Instron Co., Ltd.). The measurement was carried out in a three-point bending mode with a crosshead speed of 1.2 mm/min and support span of 48 mm according to ASTM D790. The samples used for testing are 25 mm × 6 mm × 3 mm.

3. Result and discussion

Figure 1. shows the FTIR spectra of bamboo charcoal at different residual lignocellulose content. Bamboo charcoal with higher residual lignocellulose content (BC10-20% and BC20-30%) exhibited peaks corresponding to hydroxyl (-OH) and carboxylic or carbonyl (C=O) functional groups at around 3420 and 1735 cm^{-1}, respectively. In the case of BC20-30%, both peaks was more apparent due to higher amount of residual lignocellulose content, while, BC0–10% had a significant decrease in the height of the -OH and C=O peaks due to increasing the firing temperature above 400 °C, causing the decomposition of -OH and C=O to become conjugated with aromatic rings [8].

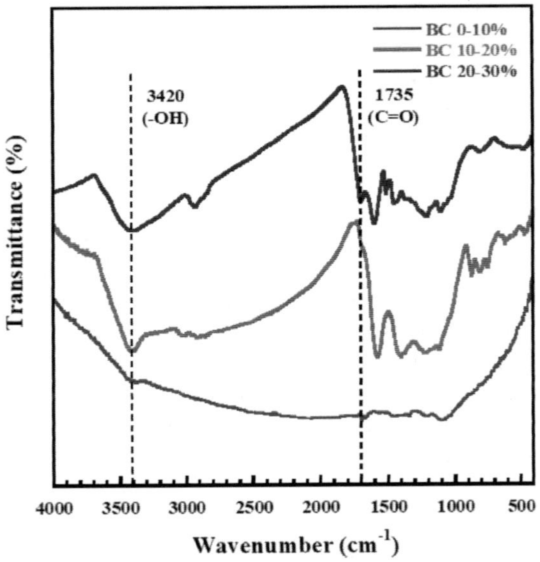

Figure 1. FTIR Spectra of bamboo charcoal at different residual lignocellulose content.

Figure 2. shows the flexural strength and flexural modulus of PFMs filled with bamboo charcoal (BC) and synthetic graphite. Flexural strength and flexural modulus of PFMs filled with BC0-10%, BC10-20%, BC20-30% was 45.5, 52.8, and 58.3 MPa, and 7.2, 8.3, and 9.1 GPa, respectively. While flexural strength and flexural modulus of PFMs filled with synthetic graphite was 48.8 MPa and 6.9 GPa, respectively. It was found that the flexural strength and the flexural modulus of PFMs filled with BC were higher than that of PFMs filled with synthetic graphite. In addition, the flexural strength and the flexural modulus of PFMs filled with BC increased with higher amount of residual lignocellulose content in bamboo charcoal. The increase in flexural strength and flexural modulus is due to the presence of residual hydroxyl and carboxylic or carbonyl groups in the lignocellulose forming hydrogen bonds with the hydroxyl groups of polybenzoxazine. This creates a strong adhesion between filler and matrix, resulting in better properties [9].

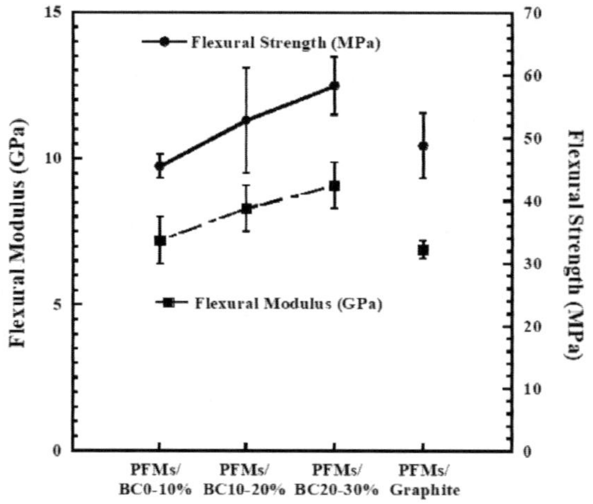

Figure 2. Flexural property of PFMs filled with bamboo charcoal and synthetic graphite.

Figure 3. shows the TGA Thermograms of PFMs filled with bamboo charcoal and synthetic graphite. Degradation temperatures at 5 % weight loss (T_{d5}) of PFMs filled with BC0-10%, BC10-20%, BC20-30% and synthetic graphite were found to be 403, 399, 385, and 408 °C, respectively. Whereas, residual weight at 1000 °C of PFMs were 71.1, 70.2, 69.3, and 72.3 %, with the addition of BC0-10%, BC10-20%, BC20-30%, and synthetic graphite, respectively. PFMs/BC0-10% showed the T_{d5} and residual weight closest to PFMs/synthetic graphite, as BC0-10% had the lowest lignocellulose content, and the functional groups had decomposed due to the carbonization process, resulting in BC0-10% being able to withstand more heat than BC10-20% and BC20-30%. The decrease in T_{d5} of PFMs/BC10-20% and PFMs/BC20-30% were observed. This behaviour may be due to higher residual lignocellulose content, having a degradation temperature range of 370-550 °C [10]. Therefore, the presence of higher amounts of lignocellulose in bamboo charcoal corresponds to lower degradation temperatures.

In addition, the deterioration of the brake pad can be caused by heat and pressure generated during braking based on the Thailand Industrial Standard (TIS: 97-2014) testing at high temperature (100 - 350 °C). Therefore, the lifespan of PFMs filled with bamboo charcoal might also be estimated by its thermal stability of the composite. From the results, thermal stability and mechanical properties of the PFMs filled with bamboo charcoal could be comparable with PFMs filled with synthetic graphite. It is acceptable that the PFMs filled with bamboo charcoal have a potential to be used as a brake pad.

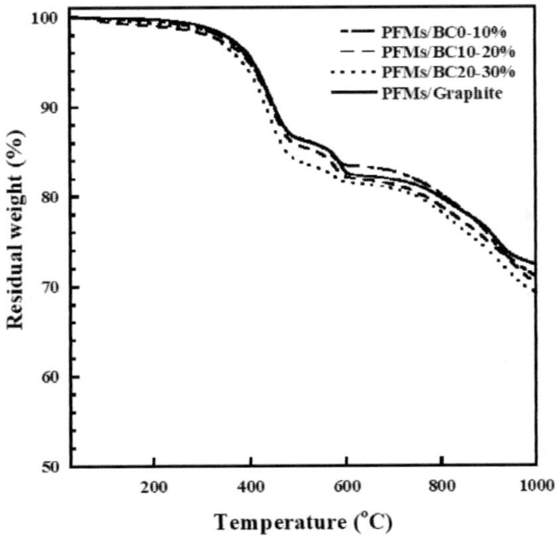

Figure 3. TGA thermograms of PFMs filled with bamboo charcoal and synthetic graphite.

4. Conclusions

In this research, the effect of lignocellulose content-containing bamboo charcoal on thermal and mechanical properties of PFMs was studied. According to the mechanical properties, PFMs/BC20-30% showed the greatest flexural strength and flexural modulus values due to the highest residual lignocellulose content. Conversely, the high residual lignocellulose content led to obtain lower T_{d5} of PFM/BC20-30% than that of PFMs filled with BC0-10% and BC10-20%. However, the T_{d5} of PFMs was within the specified range that can withstand the loss of thermal efficiency for brake pads usage.

Acknowledgments

This research has received funding support from Thailand Science Research and Innovation 2023 and Faculty of Engineering, Srinakharinwirot University (grant number 298/2565). The ingredients and lubricant (synthetic graphite) were kindly provided by Compact International (1994) Co., Ltd., Phetchaburi, Thailand., and Acme International (Thailand) Ltd., Bangkok, respectively.

References

[1] Liew K W and Nirmal U 2013 *Mater. Des.* **48** 25.
[2] Ria Jaafar T, Selamat M and Kasiran R 2012.
[3] Lertwassana W, Parnklang T, Mora P, Jubsilp C and Rimdusit S 2019 *Compos. B. Eng.* **177** 107280.
[4] Nandiyanto A, Hofifah S, Girsang G, Putri S, Budiman B, Triawan F and Al-Obaidi A S 2021 *Automotive Experiences* **4** 68.
[5] Sukrawan Y, Hamdani A and Mardani S A 2019 *Mater. Phys. Mech.* **42** 367.
[6] Bashir M, Qayoum A and Saleem S 2019 *Mater. Res. Express.* **6** 115551.
[7] Ishida H 2011 *Handbook of Benzoxazine Resins,* ed H Ishida and T Agag (Amsterdam: Elsevier) pp 3-81.
[8] Song-lin Z, Shang-yu G, Xi-gen Y and Bo-sen X 2003 *J. For. Res.* **14** 75.
[9] Kumar R, Gunjal J and Chauhan S 2021 *Compos. B. Eng.* **217** 108846
[10] Kim P, Johnson A, Edmunds C W, Radosevich M, Vogt F, Rials T G and Labbé N 2011 *Energy. Fuels.* **25** 4693.

IOP Conf. Series: Materials Science and Engineering 1280 (2023) 012006 doi:10.1088/1757-899X/1280/1/012006

Transparent flexible fluorescent films based on natural rubber composited with quantum dots for traffic equipment

O Suntijitrungruang[1,*], W Pon-On[1] and S Boonchui[1,2]

[1] Department of Physics, Faculty of Science, Kasetsart University, Bangkok, Thailand
[2] Center of rubber and polymer materials in agriculture and industry (RPM), Faculty of Science, Kasetsart University, Bangkok, Thailand

*Corresponding author: monorol@hotmail.com

Abstract. Quantum dots are one of the nanotechnology materials applied in various ways like flexible electronic devices and others. In this work, we study and synthesize semiconductor quantum dots (SQDs) to produce fluorescent natural rubber films by the technique called directed blending. We illustrate the appropriate ratio of mixtures employed in our method such as quantum dot colloidal solutions and natural latex as well as moles between each substance. During the process, the liquid composite and rubber adhesive is applied to the surface for creating flexible fluorescent films that can be dried at room temperature. In addition, these flexible fluorescent films can be covered with diverse curved surfaces applied probably as a traffic device like luminous traffic cones or traffic signs that glow at night. Based on the unique properties of our products, our research provides an application developed from quantum dot nanotechnology.

1. Introduction

The information from department of highways in Thailand describes that there are 1,024 times of road accident in May 2020, resulting in severe damages both on people's life and country's economics. To avoid harmful accidents on roads particularly in the night, it is necessary to develop fluorescent traffic equipment. Accordingly, flexible electronic devices are a key to cope with this problem. Today, there is development and various research on organic light-emitting diodes (OLED) [1-3] which are flexible [4-5] and somewhat convenient for attaching to the surface. Moreover, OLED are very efficient to alter the light from a certain wavelength to other wavelengths [6]. On the other hand, quantum dots (QDs) [7] are a type of nanotechnology studied diversely such as hybridization [8], gas sensors [9] and medical imaging [10]. Modified QDs, like OLED, demonstrate the property to shift incident light wavelengths [11]. This attribute can be determined by adjusting the size and spread of QDs, hence altering the energy state levels. On this ground, when modified QDs blend with particular latex, it probably can create promising flexible electronic devices possessing fluorescent property. In addition, latex, an emulsion (stable dispersion) of polymer microparticles in water [12], are drawn off from Pará rubber trees (*Hevea Brasiliensis*) which are planted abundantly in Thailand [13]. For this reason, to make the flexible electronic device from Pará rubber trees will definitely benefit Thai orchardists and economics. In our research, we practically create fluorescent natural rubber films from semiconductor quantum dots (SQDs) that can produce various useful luminous traffic equipment and other fluorescent devices.

IOP Conf. Series: Materials Science and Engineering 1280 (2023) 012006 doi:10.1088/1757-899X/1280/1/012006

2. Experimental

To create flexible fluorescent films according to our way, there are two significant steps that are i) synthesizing the modified QDs that consist of carbon quantum dots (CQDs) mingling with SQDs (Cd-Se) and ii) creating the prototype of the product originating from modified QDs and the specific latex. Concerning CQDs, we synthesize them by the main process referred to as hydrothermal. Regarding SQDs (Cd-Se) synthesis, we employ a method called hot injection and decomposition. The characteristic of the fluorescence relies on kinds of the solvent in the hydrothermal process, and the time duration of crystallization during SQDs (Cd-Se) synthesis. Such property is contributed to the quantum size effect.

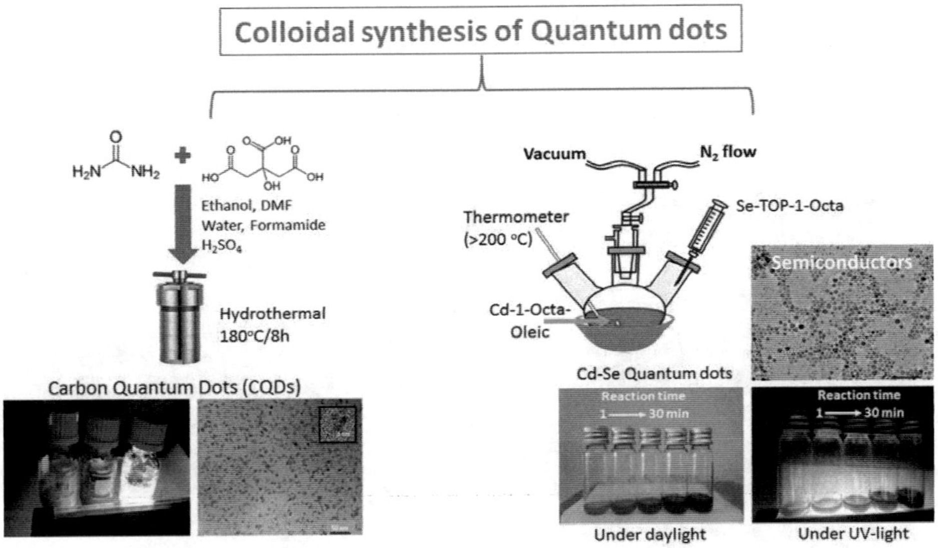

Figure 1. The process for synthesis modified QDs.

After deriving the modified QDs, we mix these QDs with dimethyl formamide (DMF) and polyvinylpyrrolidone (PVP) where it provides the colloid. Next, we utilize the process named directed blending for the colloid and natural rubber latex in order to obtain the liquid composite. We found that the appropriate component of the colloid and latex in our study is 1:1 wt/wt. The liquid composite from our method, then, coat on the sheets that can be dried at room temperature. Eventually, the flexible fluorescent films are available. These films can be applied to traffic equipment and the other surfaces.

In detail, SQDs in our study is a phosphor which is strontium aluminate mingling with lanthanide (Eu^{2+}/Dy^{3+}) ($Sr_{0.7}Al_2O_4:Eu_{0.1}^{2+}/Dy_{0.2}^{3+}$) through the solid state reaction. The substrates in the research are composed of strontium carbonate ($SrCO_3$), aluminium oxide (Al_2O_3), europium oxide (Eu_2O_3) and dysprosium oxide (Dy_2O_3) with the mole ratio 0.7: 2.0: 0.1: 0.2 of Sr, Al, Eu and Dy, respectively. This mixture is ground with the solvent ethanal, and then is burned at the temperature 1,100 °C for 6 h. Finally, we derive the phosphor whose color is shown in Figure 2. This phosphor, in addition, would be coated by silicon-carbon by specific procedures before blending the latex.

3. Results and discussion

Considering our refined phosphor by scanning electron microscope (SEM) as shown in Figure 3, the phosphor particle shape resembles general stones with a rough surface. The side of the particle is 20-30 μm approximately. In regard to the component probed by an energy dispersive spectrometer (EDS), the main elements are strontium (Sr), aluminium (Al) and Oxygen (O).

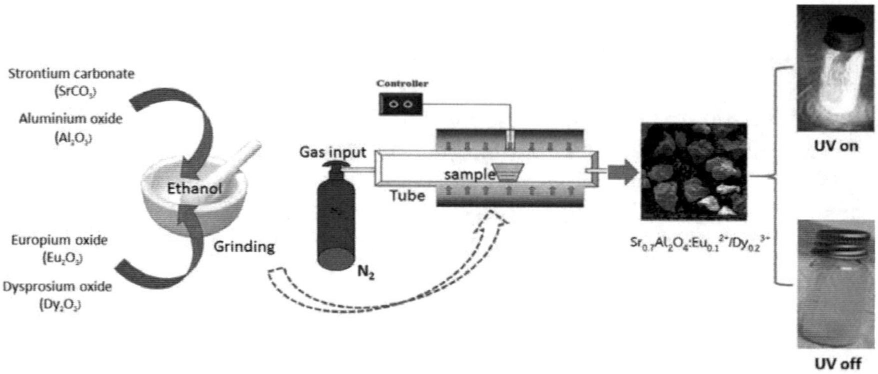

Figure 2. The process of synthesis: the phosphor (strontium aluminate) mingling with lanthanide (Eu^{2+}/Dy^{3+}) ($Sr_{0.7}Al_2O_4$:$Eu_{0.1}^{2+}$/$Dy_{0.2}^{3+}$) through the solid state reaction.

Figure 3. The characteristic of refined phosphor by SEM and the elemental component by EDS.

Considering the fluorescence spectrum of the phosphor coated with silicon-carbon, the coated phosphor emits the light with the wavelength 450 nm roughly. This wavelength is of the green color band. The intensity of each wavelength and fluorescent color are illustrated in Figure 4.

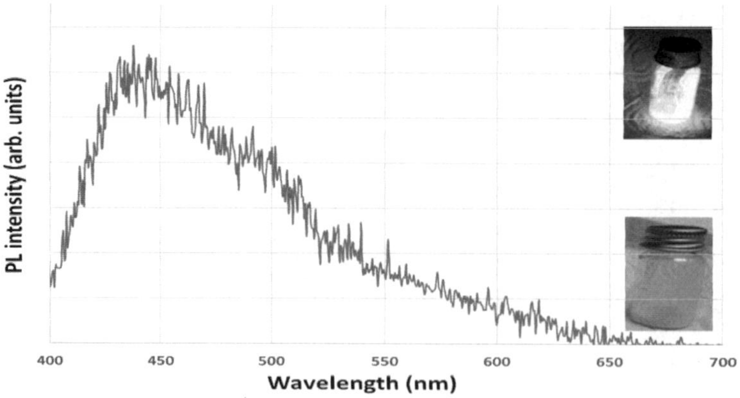

Figure 4. Spectrum of the light emitting from the phosphor coated by silicon-carbon.

After blending the liquid composite with natural rubber latex in appropriate proportion through the certain process, we derive flexible fluorescent films that can be applied to various surfaces as well as traffic equipment. Figure 5 illustrates the comparison between normal and luminous traffic cones both in the day and night.

Figure 5. The comparison between normal and luminous luminous traffic cones in the day and night.

4. Conclusions

From our work, we provide a practical way for producing the flexible electronic device being fluorescent. The method in our research employs the certain properties of modified QDs that properly blend with natural rubber latex, resulting in transparent flexible fluorescent films. The refined phosphor utilized in this study can alter the incident light and emit the fluorescent light at specific wavelengths as demonstrated in the results. The films from our research can be applied to surfaces of diverse materials especially the traffic equipment, billboards, or medical supplies, owing to its flexibility property. Our research, therefore, suggests an example of QD nanotechnology for decreasing harmful traffic accidents.

Acknowledgement
This research project was supported by National Research Council of Thailand (NRCT) and Kasetsart University Research and Development Institute (KURDI) under the grant number FF(KU)8.66.

References

[1] Zhang H, Liu X, Gong Y, Yu T and Zhao Y 2021 *Dyes Pigm.* **185**, 108969.
[2] Kamtekar K T, Monkman A P and Bryce M R 2010 *Adv. Mater.* **22** 572.
[3] Chen S, Deng L, Xie J, Peng L, Xie L, Fan Q and Huang W 2010 *Adv. Mater.* **22** 5227.
[4] National Research Council 2015 *The National Academies Press*, 105
[5] Gustafsson G, Cao Y, Treacy G M, Klavetter F, Colaneri N and Heeger A J 1992 *Nature.* **357** 477.
[6] Daskalakis KS, Freire-Fernández F, Moilanen AJ, Van Dijken S and Törmä P 2019. *ACS photonics.* **6** 2655.
[7] Valizadeh A, Mikaeili H, Samiei M, Farkhani S M, Zarghami N, Akbarzadeh A and Davaran S 2012. *Nanoscale res. Lett.* **7** 480.
[8] Cui J, Panfil Y E, Koley S, Shamalia D, Waiskopf N, Remennik S and Banin U 2019 *Nat. commun.* **10**, 5401.
[9] Suntijitrungruang O, Lakronwat J, Uthailiang T, Pongkitiwanichakul P and Boonchui S 2022 *Front. Chem.* **10** 1036197.
[10] Ramírez H Y, Flórez J and Camacho ÁS 2015 *Phys. Chem. Chem. Phys.* **17** 23938.
[11] Umesh B, Eraiah B, Nagabhushana H, Sharma SC, Sunitha DV, Nagabhushana BM, and Chakradhar RPS 2013 *Mater. Res. Bull.* **48** 180.
[12] Wang H, Yang L and Rempel G L 2013 *Polym. Reviews.* **53** 192.
[13] https://top5ofanything.com/list/7738a992/Rubber-Producing-Countries; Jan 2014.

IOP Conf. Series: Materials Science and Engineering 1280 (2023) 012007 doi:10.1088/1757-899X/1280/1/012007

Synthesis and characterization of calcium carbonate/carbon from water lettuce via hydrothermal carbonization process

N Chaiammart[1,2], M M Thu[1,3], N Srisuk[1], N Mueanpun[1] and G Panomsuwan[1,2,3,*]

[1]Department of Materials Engineering, Faculty of Engineering, Kasetsart University, Bangkok 10900, Thailand
[2]Special Research Unit for Biomass Conversion Technology for Energy and Environmental Materials, Kasetsart University, Bangkok 10900, Thailand
[3]International Collaborative Education Program for Materials Technology, Education, and Research (ICE-Matter), ASEAN University Network/Southeast Asia Engineering Education Development Network (AUN/SEED-Net), Faculty of Engineering, Kasetsart University, Bangkok 10900, Thailand

*Corresponding author: gasidit.p@ku.ac.th

Abstract. The conversion of biomass into value-added products has recently received much attention for a broad range of applications. In this work, water lettuce was converted into calcium carbonate ($CaCO_3$)/carbon via hydrothermal carbonization (HTC) and pyrolysis at 900 °C. The HTC temperature and time varied in the range of 160–200 °C for 6–18 h, respectively. X-ray diffraction analysis indicated that the samples consisted of a mixture of calcite and vaterite phases of $CaCO_3$ and amorphous carbon. The ratio of calcite and vaterite phases varied with HTC time. The Fourier transform infrared spectroscopy (FTIR) result showed the characteristic absorption bands confirming the presence of $CaCO_3$. Scanning electron microscopy (SEM) images revealed the large crystal of $CaCO_3$ and fine carbon particles. From the N_2 sorption analysis, the sample prepared from the HTC at 200 °C for 6 h had the highest specific surface area of 95 m^2/g due to the development of micropores. The results presented in this work demonstrated that both HTC temperature and time play critical roles in altering the surface area and phase structure of $CaCO_3$/carbon. The $CaCO_3$/carbon derived from water lettuce can potentially be used and adapted for many applications.

1. Introduction

Water lettuce (*Pistia Stratiotes L.*) is a native aquatic plant that can float on the water surface. It is known as invasive alien species due to its high survivability and resistivity in wide environmental conditions [1]. Moreover, its rapid overgrowth by sprouting in water causes ubiquitous environmental problems and serious ecological damages, such as blocking waterways, reducing native species abundance, altering aquatic life, reducing oxygen levels, and blocking the air/light shining–water interface [2]. Therefore, efficient management and utilization of excess water lettuce are highly needed to solve problems and create value for society.

Recently, the utilization of waste biomass has relied on conversion into value-added products via economic and low-cost methods [3]. By considering the constituent of water lettuce, it is mainly composed of the lignocellulosic structure, which contains lignin, hemicellulose, and cellulose. These are hydrocarbon-based materials that can be used as sources for the synthesis of carbonaceous materials [4]. Additionally, several metal ions (e.g., Mg^{2+}, Ca^{2+}, Zn^{2+}, Si^{4+}, etc.) were also detected in water lettuce due to their good absorptivity [5]. The type and amount of metal ions varied broadly, depending on the environmental conditions of water resources. These metal ions may lead to the formation of several inorganic compounds during the conversion processes [6].

The purpose of this work is to convert water lettuce into value-added products via hydrothermal carbonization (HTC) and pyrolysis processes. HTC temperature and time were varied in the range of 160–200 °C and 6–18 h, respectively, while the pyrolysis temperature was kept at 900 °C. The effect of HTC temperature and time on the phase, chemical functionality, specific surface area, and morphology of the resulting products were investigated by X-ray diffraction (XRD), Fourier transform infrared spectroscopy (FTIR), N_2 sorption analysis, and scanning electron microscopy (SEM), respectively.

2. Experimental

2.1. Preparation of $CaCO_3$/carbon from water lettuce

Water lettuce was washed with water and dried at 105 °C for 24 h in an oven. The dried water lettuce was crushed into a fine powder and then sieved with a 0.79 mm-screen mesh. Next, 3 grams of water lettuce powder was mixed with 45 mL of deionized water and loaded into a Teflon-lined stainless-steel autoclave. The autoclave was then heated in an oven for the HTC process at the temperature range of 160–200 °C, while HTC time varied from 6 to 18 h. After cooling, the samples were washed with deionized water and then dried in the oven at 105 °C for 24 h. The samples were loaded into a ceramic boat and pyrolyzed in a tube furnace at 900 °C for 1 h with a heating rate of 10 °C/min under the argon flow. The obtained samples were named WL-*x*-*y*, where *x* and *y* represent HTC temperature and time, respectively. The yields of the WL-*x*-*y* samples varied in the range of approximately 12–18% with respect to the weight of the starting dried water lettuce.

2.2. Characterization

The phase structure was identified using an X-ray diffractometer (PANalytical, Empyrean). The specific surface area and porosity were evaluated using a surface area analyzer (Micromeritics, 3Flex) at a temperature of –196 °C. The degassing process was performed at 150 °C for 12 h before measurement. The chemical bonding was identified using a Fourier transform infrared spectrometer (Bruker, Alpha II). Morphology and elemental analysis were conducted using a scanning electron microscope (HITACHI, SU3500).

3. Results and discussion

The XRD patterns are displayed in Figure 1a. The diffraction peaks at 23.0°, 29.4°, 36.0°, 39.4°, 43.0°, and 47.6° can be assigned to the (012), (104), (110), (113), (202), and (024) planes of the calcite phase of $CaCO_3$, respectively. The diffraction peaks of the vaterite phase of $CaCO_3$ were also observed at 20.9°, 24.9°, 27.0°, 32.7°, 43.9°, and 50.0°, corresponding to the (002), (100), (101), (102), (110), and (104), respectively. These results indicate that the presence of Ca^{2+} ions in water lettuce reacted with the carbon and oxygen from the lignocellulosic structure of water lettuce, giving rise to the formation of $CaCO_3$. With increasing HTC temperature, no obvious change in the vaterite phase was seen since the calcite phases do not have enough time to transform into the vaterite phase during the HTC process. Meanwhile, the vaterite phase was changed with the HTC time at 200 °C. The vaterite phase increased and reached maximum intensity for WL-200-12 but decreased for WL-200-18. In addition, a weak and broad peak feature corresponding to the amorphous carbon was also present at around 23°–25°, as shown in the magnified XRD pattern at the low-intensity region in Figure 1b.

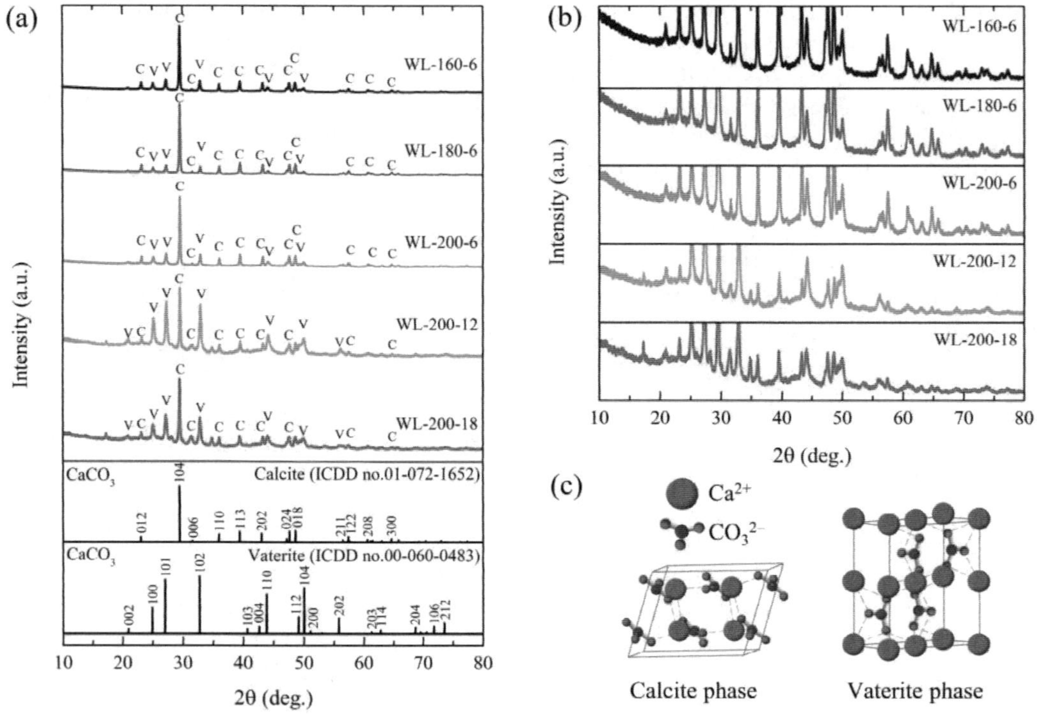

Figure 1. (a) XRD patterns, (b) magnified XRD patterns at the low-intensity region, and (c) crystal structures of calcite and vaterite phases of $CaCO_3$.

The FTIR spectra of all samples are shown in Figure 2a. The absorption peak at 1428 cm^{-1} and 873 cm^{-1} can be assigned to the vibration modes of the C–O bending in the carbonate group (CO_3^{2-}), which are the characteristics of $CaCO_3$ [7]. As the HTC temperature increased, the peak at 1428 cm^{-1} became more intense, which is due to the transformation from calcite to vaterite phases [8]. Additionally, the peak at 1112 cm^{-1} and 1033 cm^{-1} are associated with the C–O stretching mode of carbons and suppressed at high HTC temperatures and times [9].

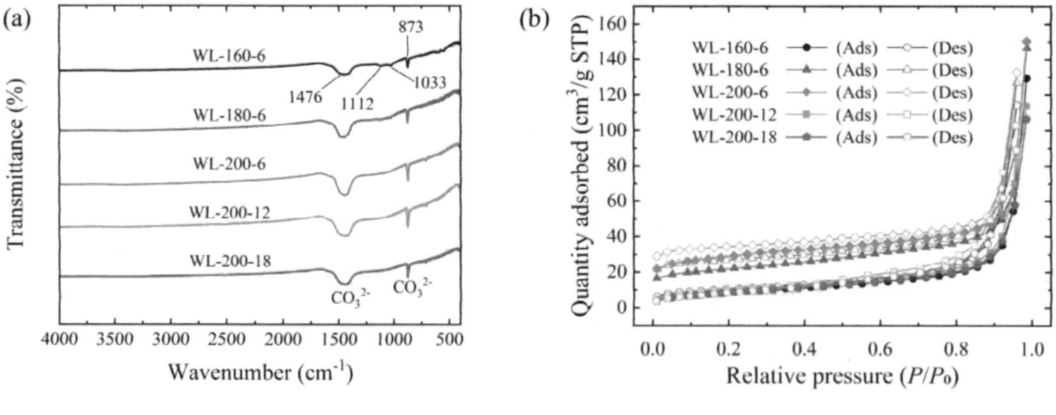

Figure 2. (a) FTIR spectra and (b) N_2 adsorption-desorption isotherms of all samples.

Furthermore, the surface area and porosity of all samples were evaluated from the N_2 adsorption-desorption isotherms in Figure 2b. All samples mainly exhibited a type II isotherm with an abrupt increase in N_2 uptake at high relative pressures, indicating a non-porous structure with the major presence of macropores. The specific surface area was determined from adsorption isotherm (P/P_0 = 0.05–0.30) using Brunauer–Emmett–Teller (BET) theory. With increasing HTC temperature, the specific surface area significantly increased from 31 m^2/g for WL-160-6 to 95 m^2/g for WL-160-18. An increased specific surface area was contributed by the development of micropores, as indicated by an increase in quantity adsorbed at low relative pressure. However, prolonging HTC time led to a decrease in specific surface area due to the deterioration of micropores. The authors believe that the specific surface area of samples was mainly contributed by carbons rather than $CaCO_3$. Textural parameters obtained from the N_2 sorption analysis are listed in Table 1.

Table 1. Textural parameters obtained from the N_2 adsorption-desorption analysis of all samples.

Sample	BET surface area (m^2/g)	Micropore area (m^2/g)	External surface area (m^2/g)	Pore volume (cm^3/g)	Micropore volume (cm^3/g)
WL-160-6	31	2 (6%)	29 (94%)	0.158	0.001
WL-180-6	72	29 (40%)	43 (60%)	0.175	0.015
WL-200-6	95	36 (37%)	59 (63%)	0.173	0.019
WL-200-12	35	0 (0%)	35 (100%)	0.151	0
WL-200-18	32	0 (0%)	32 (100%)	0.126	0

The SEM images of all samples in Figure 3 present the mixture of a large crystal of $CaCO_3$ and fine-particulate carbons. When the HTC temperature increased (Figures 3a–3c), the carbon particles (indicated as yellow arrows) became finer due to microporous development. However, at longer HTC duration at 200 °C, the number of fine carbon particles reduced significantly, while the rod-like shape of calcite (indicated as blue arrows) and spherical particles of vaterite (indicated as green circles) began to be observed (Figures 3d and 3e) [10, 11]. Figure 4 shows the representative EDS mapping images of WL-180-6 and its corresponding spectrum. The clear distribution of C, O, and Ca elements throughout the sample again confirms the presence of carbon and $CaCO_3$. The morphological evolution of the samples is well consistent with the above results obtained from the XRD, FTIR, and surface area analyses.

Figure 3. SEM images at 3000× magnification of (a) WL-160-6, (b) WL-180-6, (c) WL-200-6, (d) WL-200-12, and (e) WL-200-18.

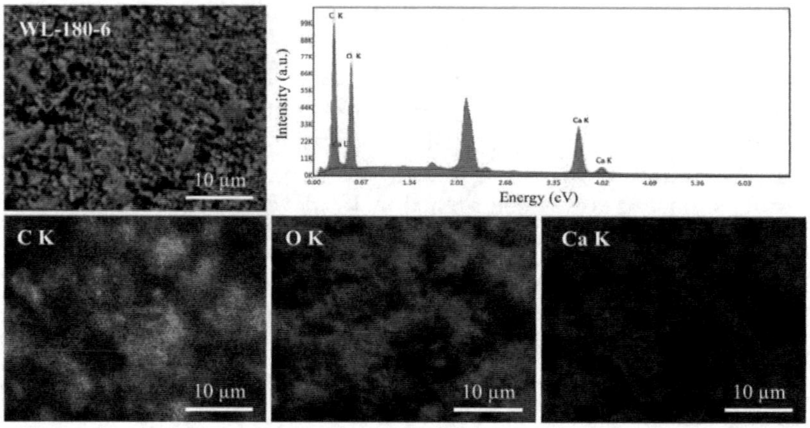

Figure 4. Representative EDS mapping images showing the distribution of C, O, and Ca elements of WL-180-6 and its corresponding spectrum.

4. Conclusions

Water lettuce was successfully converted into $CaCO_3$/carbon via HTC and pyrolysis processes. The samples contained a mixture of calcite and vaterite phases of $CaCO_3$ with amorphous carbon. The transformation from calcite to vaterite phases was observed at a longer HTC time at 200 °C. High HTC temperature could lead to more development of micropores in the carbon phase, thereby increasing the specific surface area. However, prolonging HTC time at 200 °C could deteriorate the microporous structure and lower the specific surface area. The morphology of the samples consisted of large crystals of $CaCO_3$ and fine particles of carbon. HTC temperature and time play essential roles in altering the phase structure (calcite and vaterite) and the porous structure of the samples. We anticipate that $CaCO_3$/carbon derived from water lettuce can further be applied for various energy and environmental applications.

Acknowledgement

This work was supported by the Kasetsart University Research and Development Institute (KURDI, grant no. grant no. FF (KU) 25.64) and the Research Assistant Scholarship (no. 64/09/MATE/D.Eng) from the Faculty of Engineering, Kasetsart University.

References

[1] Reddy K R and Debusk W F 1984 *Econ. Bot.* 38 229–239.
[2] Weldon LW and Blackburn R D 1987 *Weeds* **15** 5–9.
[3] Yu Z, Ma H, Liu X, Liu X, Wang M and Wang J 2022 *Green Chem Eng.* **3** 304.
[4] Varma R S 2019 *ACS Sust. Chem. Eng.* **7** 6458.
[5] Lu Q, He L, Graetz D A, Stoffella P J and Yang X 2011 *Environ. Sci. Pollut. Res.* **18**, 978–986.
[6] Iraola-Arregui I, Van Der Gryp P and Görgens J F 2018 *Waste Manage.* **79**, 667.
[7] Janbooranapinij K, Yimponpipatpol A, Ngamthanacom N, Suthiprapar J and Panomsuwan G 2021 *Clean. Eng. Technol.* **4** 100150.
[8] Siva T, Muralidharan S, Sathiyanarayanan S, Manikandan E and Jayachandran M 2017 *J. Inorg. Organomet. Polym. Mater.* **27** 770.
[9] Sanito R C, Lidwina C, Yang H H and Wang Y F 2022 *Heliyon.* **8** e11240.
[10] Konopacka-Łyskawa D 2019 *Crystals.* **9** 223.
[11] Ševčík R, Šašek P, Viani, A 2018 *J. Mater. Sci.* **53** 4022.

IOP Conf. Series: Materials Science and Engineering 1280 (2023) 012008 doi:10.1088/1757-899X/1280/1/012008

Bacterial cellulose: An eco-friendly low cost biomaterial for dye removal

S Sukphan, P Buapho, N Laingaumnuay and P Jaturapiree*

Department of Biotechnology, Faculty of Engineering and Industrial Technology, Silpakorn University, Muang, Nakhon Pathom, 73000, Thailand.

*Corresponding author: jaturapiree_p@su.ac.th

Abstract. Bacteria cellulose (BC) is a non-toxic, environmentally friendly, and biodegradable material. Additionally, it is a cheap, easily accessible, and renewable resource. It possesses exceptional properties and is widely used in a variety of industries, including textiles, paper manufacturing, food, pharmaceutical, and waste treatment. The purpose of this study was to assess the capacity for bioadsorption of pure BC in its wet and dry forms . After 210 minutes of incubation at 37°C, 120 rpm, wet and dry BC exhibit the maximum decolorization of Crystal violet (100 mg/L) at 62.72% and 71.32%, and Congo red (100 mg/L) at 44.28% and 35.23%, respectively. Additionally, the characterization and kinetic model of BC were investigated in parallel. In summary, the findings of this work imply that the application of BC in bioadsorption of dyes may be a potential and successful method for solving the concerns of environmental pollution in a way that is low-cost, simple, and environmentally friendly.

1. Introduction

Problems with wastewater from the textile sector polluting the aquatic ecosystem must be addressed immediately. One of the major causes of water pollution is the untreated discharge of industrial dye-containing effluents into the water. This is because the majority of dyes used in the textile industry have complex structures that are difficult to degrade. As a result, it will prevent the light from penetrating the water. The rate of photosynthesis is slowed down when light is blocked, which has an impact on the growth of aquatic and land creatures as well as the food chain [1, 2, 3]. Additionally, some dyes have the potential to be carcinogenic, mutagenic, and genotoxic. Methods for dye removal that have been reported include membrane filtration, ozone treatment, chemical coagulation, reverse osmosis, ion exchange, photochemical treatment, biological treatment, and adsorption [4, 5, 6]. Among all these methods, adsorption is a cost-effective, efficient, and simple form of wastewater treatment. Bacterial cellulose (BC) is a high-purity purity natural biopolymer with superior properties. It possesses a unique surface area and three-dimensional pores that allow for molecular transport. It can also absorb or bond to water. Furthermore, it is inexpensive, biodegradable, non-toxic, and ecologically beneficial. It is also a widely available and renewable resource [7, 8, 9]. This study compared the adsorption abilities of pure BC in its dry and wet forms. Additionally, a parallel investigation of BC's characterization and kinetic model was conducted.

2. Experimental

Production of BC

The bacterial strain used for the production of BC was *Acetobacter xylinum* TISTR 975. It was grown in a medium composed of 1,000 mL of filtered coconut water, 50 g of sucrose, 5 g of ammonium sulfate, and 1% (v/v) acetic acid, and the liquid culture was incubated under the static condition at room temperature (28°C–30°C) for 4 d until the BC pellicles appeared to be 0.5 cm thick. The BC pellicles were then collected and boiled in 0.5% NaOH for 30 mins. Following that, the pure BC pellicles were rinsed with running water until the pH level approximated a neutral state.

Preparation of wet and dry BC

Some of the pure BC pellicles were stored in distilled water in the refrigerator at 4°C and used as wet BC. The others were dried at 50 °C for 8-12 h before being used as dry BC. Following that, both forms of BC were sliced to 1x1 cm size to use in the dye absorption process.

Decolorization of dye by wet and dry BC

Crystal violet (CV) and Congo red (CR) dye solution at concentrations of 100 mg/L were prepared. Ten pieces of BC were placed in 25 mL of each dye solution and incubated for 120 mins under shaking at 120 rpm and 37 °C. Control samples were done in parallel without BC. To measure the residual dye, aliquot samples were taken at various time intervals and analyzed using spectrophotometry at their maximum wavelength (CV at 590 nm and CR at 490 nm., respectively). Dye removal was calculated in terms of percentage as follows,

$$\% \text{ Decolorization} = \frac{A_0\text{-}A_t}{A_0} \times 100 \quad (1)$$

where A_0 is the absorbance of the control and A_t is the absorbance of the test sample.

Characterization of BC

Dry BC samples were analyzed using X-ray diffraction (XRD) and a Fourier transform infrared spectrometer (FTIR) at the Scientific and Technological Equipment Centre, Faculty of Science, Silpakorn University.

Adsorption kinetics

The samples were taken at various time intervals, and the dye concentration was quantified using a spectrophotometer at their maximum wavelength. Using the following equations, the pseudo-first-order and pseudo-second-order kinetics models were employed to fit the adsorption kinetics data.

pseudo-first-order kinetic:
$$\ln(q_e\text{-}q_t) = \ln(q_e)\text{-}k_1 t \quad (2)$$

pseudo-second-order kinetic:
$$\frac{t}{q_1} = \frac{1}{q_e^2 k_2} + \left(\frac{1}{q_e}\right) t \quad (3)$$

where q_e and q_t (mg/g) are the amounts of dye uptake onto the surface of BC at equilibrium and time t, k_1 (min^{-1}) and k_2 ($g\ mg^{-1}\ min^{-1}$) are defined as the rate constant of pseudo-first-order kinetic model and pseudo-first-order kinetic model, respectively.

3. Results and discussion

Firstly, wet and dry BC pellicles were obtained (Figure 1a, b), secondly, the CV and CR dye solution were incubated with 10 pieces of 1x1 cm size of wet and dry BC pellicles (Figure 1c, d). The wet and dry BCs exhibit the maximum decolorization of CV (100 mg/L) at 62.72% and 71.32% and Congo red (100 mg/L) at 44.28% and 35.23%, respectively after 210 min of incubation at 37°C and 120 rpm. Figure

2 shows images and spectrophotometric scans of the CV and CR dye solutions incubated with wet and dry BCs. Numerous studies suggested that BC can be used as a bioadsorbent to remove different dyes, including Drimarene Red [10], Reactive Blue 171, Remazol Brilliant Blue R, and Chrocion Green [11].

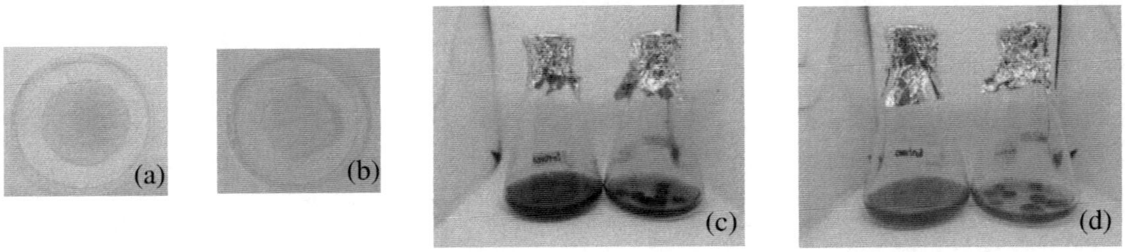

Figure 1. The photographs of the wet (a) and dry (b) BC pellicles and the CV (c) and CR (d) dye.

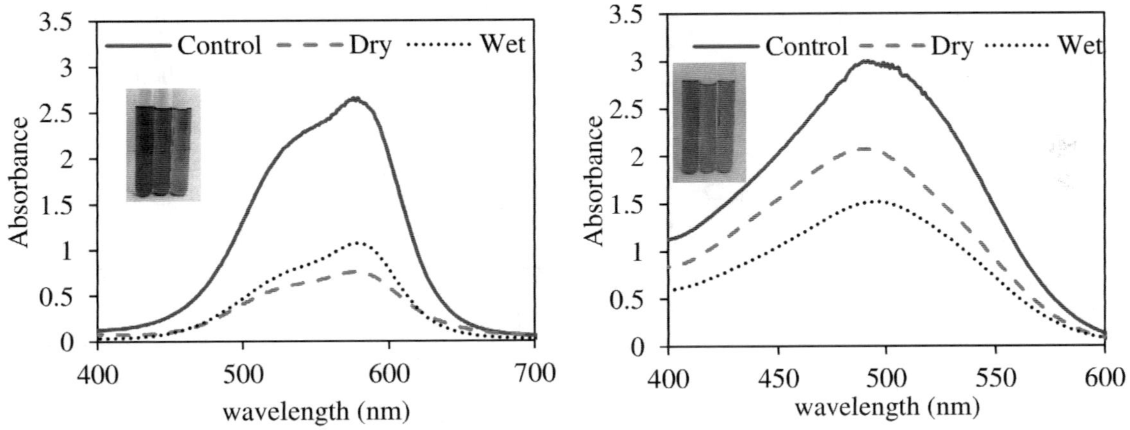

Figure 2. The photographs and spectrophotometric scans of the CV and CR dye solutions.

In the XRD patterns (Figure 3a), BC had two dominating peaks (14.57° and 23.08°), whereas CVBC (crystal violet absorbed on bacterial cellulose) (14.93°, 17.16°, and 22.97°) and CRBC (congo red absorbed on bacterial cellulose) (14.60°, 17.21°, and 22.89°) had three main characteristic peaks. Bacterial cellulose is a semicrystalline substance comprising crystalline and amorphous areas. The existence of type I and type II cellulose may be determined using the two main peaks obtained in the analyses [12]. Figure 3b shows the FTIR spectrum for BC, CVBC, and CRBC. The spectrum showed several important peaks wavenumber at 3345 cm^{-1} (hydroxyl group (-OH)), at 2921 cm^{-1} (methyl group (-CH)), at 1635-1645 cm^{-1} (C=O), at 1160 cm^{-1} (C–O–C) and at 1108-1059 cm^{-1} (oscillating C–O). This finding suggested that dye had no impact on the structure of bacterial cellulose. It might be reused for the next adsorption cycle to remove dye from the aqueous solution.

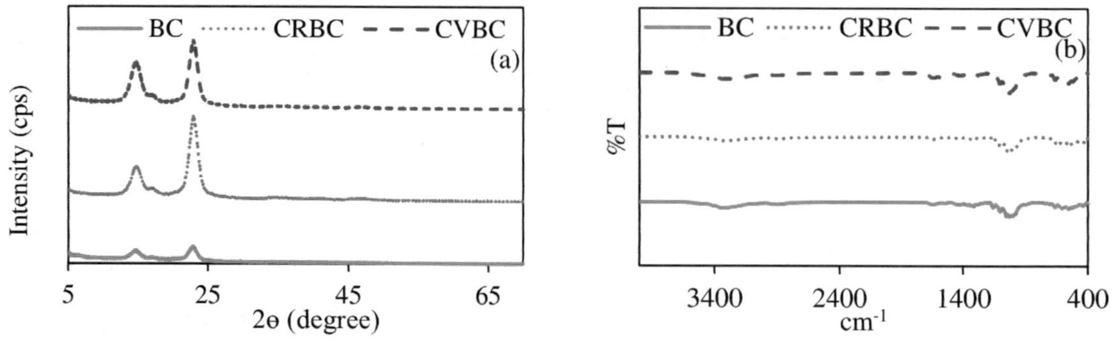

Figure 3. X-ray diffractograms (a) and FTIR chromatogram (b) of BC (blue line), CVBC (violet line), and CRBC (red line).

Table 1 summarizes the constant values for each model together with the corresponding correlation coefficient, indicating how well each theoretical model fits the experimental data. The correlation coefficient revealed that the pseudo-second order model achieved a satisfactory fit to the experimental data. Prior studies have also reported similar dye adsorption kinetics patterns [13].

Table 1. Kinetics model constants for the CR and CV adsorption on BC.

	CR (100 mg/L)		**CV** (100 mg/L)	
	Wet BC	Dry BC	Wet BC	Dry BC
q_e,exp (mg g^{-1})	0.24	8.82	0.19	42.39
Pseudo first order				
R^2	0.9639	0.9518	0.9575	0.9602
k_1 (min^{-1})	0.0149	0.0141	0.0149	0.0141
q_e,cal (mg g^{-1})	0.40	0.74	0.51	1.19
Pseudo second order				
R^2	0.9553	0.9703	0.9862	0.9993
k_2 (g min^{-1} mg^{-1})	0.1092	0.0094	0.2152	0.0044
q_e,cal (mg g^{-1})	0.26	6.00	0.21	43.48

4. Conclusions

In this research, wet and dry BC were used as an adsorbent for the removal of CV and CR dye solutions. We discovered that the use of BC as the bioadsorption of dyes may be an effective and potentially low-cost, easy, and green strategy to address environmental pollution concerns.

Acknowledgment

This research is financially supported by Thailand Science Research and Innovation (TSRI) National Science, Research and Innovation Fund (NSRF) (Fiscal Year 2023) and the Department of Biotechnology, Faculty of Engineering and Industrial Technology, Silpakorn University.

References

[1] Jahan K, Tyeb S, Kumar N and Verma V 2021 *J. Polym. Environ.* **29** 1257.
[2] Mohite B V and Patil S V 2014 *J. Biomater. Sci. Polym. Ed.* **25** 2053.
[3] Kamal T, Ul-Islam M, Khan S B and Asiri A M 2015 *Int. J. Biol. Macromol.* **81** 584.
[4] Isik Z, Unyayar A and Dizge N 2018 *Avicenna. J. Environ. Health. Eng.* **5** 106.
[5] Li D, Li Q, Bai N, Dong H and Mao D 2017 *ACS Sustainable Chem. Eng.* **5** 5598.
[6] Leal A, Lima A, Azevedo M, Santos D, Zaidan L, Lima V and Filho I 2021 *Scientia. Plena.* **17**.

[7] Zhong C 2020 *Front. Bioeng. Biotechnol.* 8.
[8] Lahiri D, Nag M, Dutta B, Dey A, Sarkar T, Pati S, Edinur HA, Abdul Kari Z, Mohd Noor NH and Ray R R 2021 *Int. J. Mol. Sci.* **22** 12984.
[9] Ashour R M, Abdel-Magied A F, Wu Q, Olsson R T and Forsberg K 2020 *Polymers* **12** 1104.
[10] Vyjayanthi, J. P., and Sumathi Suresh. 2010 *Water. Environ. Res.* **82** 601.
[11] Filiz BORAN 2022 *Comm. J. Biol.* **6** 1.
[12] Trilokesh C and Uppuluri K B 2019 *Sci. Rep.* **9** 16709.
[13] Ai L, Jiang J and Zhang R 2010 *Synth. Met.* **160** 762.

Production of dextran by Lactobacillus fermentum OR1 isolated from pickle mango juice

N Karnthanachotikul and P Jaturapiree*

Department of Biotechnology, Faculty of Engineering and Industrial Technology, Silpakorn University, Muang, Nakhon Pathom, 73000, Thailand.

*Corresponding author: jaturapiree_p@silpakorn.edu

Abstract. Dextran is an exopolysaccharide produces from bacteria that have been shown to be useful in the pharmaceutical and food industries. The goal of this study was to screen bacteria capable of producing dextran from Thai fermented foods and then maximize dextran production. The maximal amount of dextran produced by *Lactobacillus fermentum* OR1 isolated from pickled mango juice is 65 g/L after 20 h of incubation at 30°C and 200 rpm in DMB2 Medium containing 160 g/L sucrose. Dextran's morphology, chemical structure, and thermal stability were further characterized using SEM and FTIR. The results suggest that *Lactobacillus fermentum* OR1 might be widely used in the production of linear dextran which has the potential to be used as a natural agent in food and other industries.

1. Introduction

Dextrans are produced from sucrose by dextransucrases, which are secreted into the culture medium by a variety of microorganisms including *Leuconostoc*, *Lactobacillus*, *Pediococcus*, *Lactococcus*, *Weissella*, and *Streptococcus* species. Dextran is a long-chain exopolysaccharide composed mainly of D-glucose molecules. It has a primary backbone made up of linear α-(1–6) links, while branches have varying numbers of links α-(1–2), α-(1–3), or α-(1–4). The structure of dextran, such as chain length and percentage of branching, is depended on the strain that produces it, as well as, the fermentation conditions and available nutrients. [1,2,3] Dextran is widely used in both the food and non-food industries. In the food industries, they act as cryoprotectants or moisture-increasing agents and can improve the texture, rheology, and flavour of several beverages. In other fields, they operate as an agent (Sephadex gels) in the bioseparation process, have antiviral action in salmonids, and are used as a plasma replacement, iron transporter, or anticoagulant in the pharmaceutical industries. However, the wide range of dextran uses, as well as the growing demand for the free-additive product, require the search for new dextran-producing bacteria. [4,5,6] Thus, the goal of this study was to isolate and screen dextran-producing bacteria from Thai fermented foods and then maximize dextran production. The morphology, chemical structure, and thermal stability of the dextran were also investigated.

2. Experimental

Isolation and screening of dextran-producing bacteria

Thai fermented food was purchased from the local market (Nakorn Pathom, Thailand). Pickled mustard (PC), and pickled garlic (GL) 1 g or pickled mustard juice (PCJ), pickled garlic juice (GLJ), and pickled mango juice (PMJ) 1 mL added to a test tube containing 0.85% NaCl 9 mL. The appropriately diluted samples were cultivated on agar MRS and incubated at 37 °C for 24-48 h until the single colony appeared. A single colony was placed in 5 mL of MRS medium and cultured at 30 °C for 24 h. The inoculum was then transferred to 50 mL MRS medium and incubated at 30 °C and 200 rpm for 24 h. The bacteria with the highest dextran production were chosen for future studies.

Production of dextran
The selected strain was cultured in 50 mL Dextran Medium Broth 2 (DMB2) in 250 mL Erlenmeyer flasks. The following ingredients were used to make one liter of DMB2: sucrose, 5 g becto-peptone, 5 g yeast extract, 15 g K_2HPO_4, 0.01 g $MnSO_4 \cdot H_2O$, 0.01 g NaCl, and 0.05 g $CaCl_2 \cdot 2H_2O$. The fermentation was carried out at 30°C on a rotary shaker. At time intervals, samples were taken out for further analysis. The cells were removed after centrifugation at 4 °C and 10,000 rpm for 30 min. The supernatant was mixed with a 1:1 ratio of 95% cold ethanol, which was then kept at 4 °C overnight. The dextran precipitate was collected through centrifugation at 4,000 rpm at 4 °C for 20 min. This procedure was performed twice to purify the dextran. The dextran yield was calculated by weighing the polysaccharide mass after drying it at 50 °C until it reached a consistent weight.

Characterization of dextran
Dry dextran samples were characterized at the Scientific and Technological Equipment Centre, Faculty of Science, Silpakorn University, using a scanning electron microscope (SEM) and a Fourier-transform infrared spectrometric (FTIR).

3. Results and discussion
In this study, a dextran-producing bacterial strain named *Lactobacillus fermentum* OR1 was isolated from pickled mango juice. In MRS broth, it showed a highly viscous liquid (Figure 1a). A bacterial strain was capable of forming a tiny, creamy, and slimy colony on an MRS plate containing sucrose, indicating the existence of dextran (Figure 1b). As shown in Figure 1C, the maximum dextran production was 28.76 g/L.

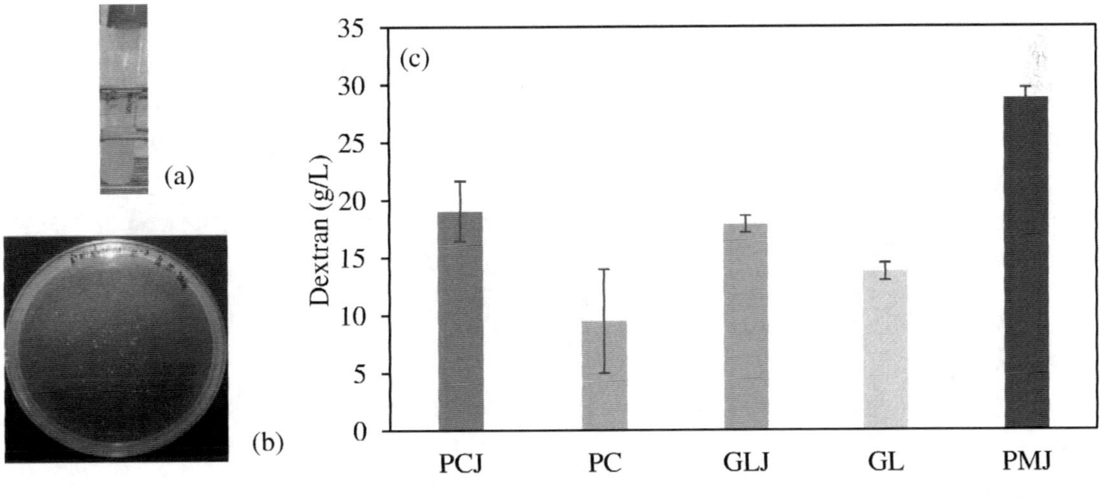

Figure 1. *Lactobacillus fermentum* OR1 in MRS broth (a) and on MRS agar (b). Dextran production by bacteria isolated from various samples (c).

The typical growth curve and dextran synthesis by *Lactobacillus fermentum* OR1 in 250 mL Erlenmeyer flasks with 50 mL of Dextran Medium Broth 2 (DMB2) were illustrated in Figure 2a. After 20 hours of incubation, a maximum dextran concentration of 35 g/L was recorded. Thus, the following experiment

IOP Conf. Series: Materials Science and Engineering 1280 (2023) 012009 doi:10.1088/1757-899X/1280/1/012009

was performed at 20 hours. Sucrose concentrations ranging from 80 to 240 g/L were used to investigate the influence of initial sucrose concentrations (Figure. 2b). The maximal dextran synthesis increased from 25 to 65 g/L as the initial sucrose level increased. According to Kim et al., this may be explained by the enhanced activity of dextransucrase in the medium [7]. At an initial concentration of 160 g/L, the highest dextran yield was 65 g/L. Dextran synthesis will be inhibited by the substrate at greater sucrose concentrations of 160 g/L.

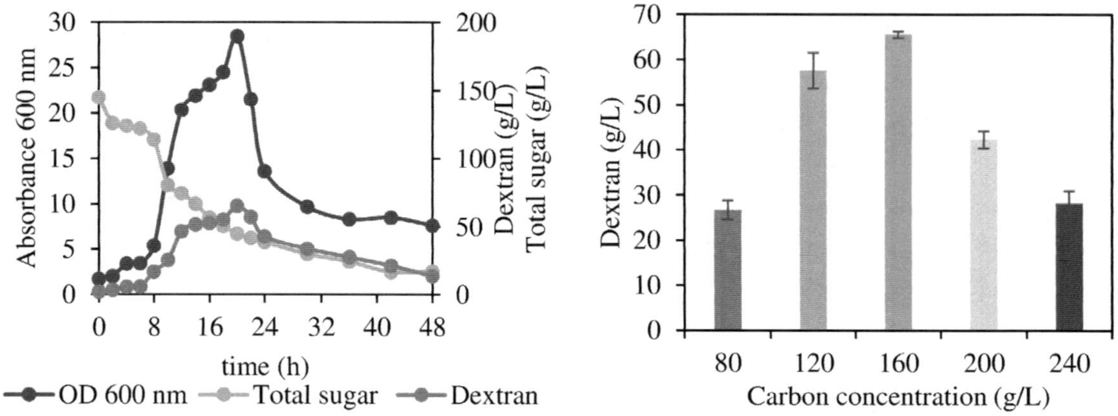

Figure 2. (a) The growth and dextran production by *Lactobacillus fermentum* OR1, (b) The effects of sucrose concentration on dextran production

From the supernatants, the dextran produced by *Lactobacillus fermentum* OR1 was purified (Figure 3a), and its structure and content were characterized. Dextran surface morphology is shown by SEM, which can also emphasize the properties of polymers. In Figures 3b and 3c, dextran's microstructure and surface morphology micrographs were presented at 200x and 2000x, respectively. The dextran had a smooth, glittering surface that was like a sheet, which is a positive quality of the materials employed to apply as the bioplastic films. In the current investigation, FTIR was employed to identify and characterize the functional groups of dextran. Dextran's spectra were investigated between 500 cm^{-1} and 4000 cm^{-1}. According to Figure 4, the peak in the ranges of 3420.46 cm^{-1} and 2925.27 cm^{-1} were related to the hydroxyl stretching and C-H stretching vibrations of the polysaccharide, respectively [8]. The presence of a linear glucose chain in α-(1-6) glycosidic bond formation was suggested by the strong signal at 1,018.73 cm^{-1} [9].

(a) (b) (c)

Figure 3. (a) purified dextran produced by *Lactobacillus fermentum* OR1 and SEM showing the surfacemorphology of purified dextran produced by *Lactobacillus fermentum* OR1 at various magnifications at 200× (b), 2000× (c)

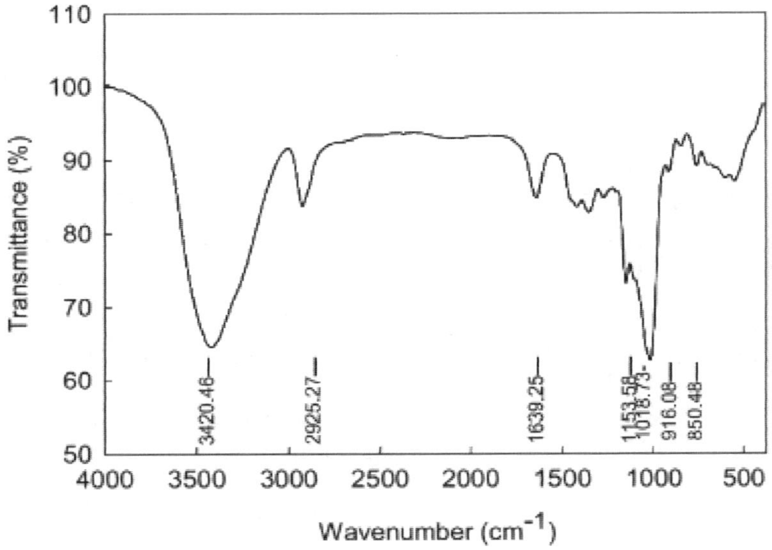

Figure 4. FT-IR spectrum of purified dextran produced by *Lactobacillus fermentum* OR1.

4. Conclusions

Screening dextran-producing strains to increase maximal dextran and useful structure and characteristics for several industrial applications is key. This study identified *Lactobacillus fermentum* OR1 as the dextran-producing strain that was isolated from pickled mango juice (PMJ). The highest dextran concentration was 65 g/L in Dextran Medium Broth 2 (DMB2), which had an initial sucrose concentration of 160 g/L. This finding implies that it is possible to apply purified dextran as a natural additive in food and other industries.

Acknowledgment

This work was supported by the Department of Biotechnology, Faculty of Engineering and Industrial Technology, Silpakorn University, and by The National Research Council of Thailand, Academic Year 2017.

References

[1] Naessens M, Cerdobbel A, Soetaert W and Vandamme E J 2005 *J. Chem. Technol. Biotechnol.* **80** 845.
[2] Xue N, Svensson B and Bai Y 2022 *Carbohydr. Polym.* **275** 118705.
[3] Díaz-Montes E 2021 *Polysaccharides* **2** 554.
[4] Sarwat F, Qader S A U, Aman A and Ahmed N 2008 *Int. J. Biol. Sci.* **4** 379.
[5] Suryawanshi N, Naik S and Jujjawarapu S E 2022 *Appl. Food Biotechnol.* **5** 22.
[6] Nabot M, Guérin M, Sivakumar D, Remize F and Garcia C 2022 *Biology* **11** 171.
[7] Kim D, Robyt J F, Lee S Y, Lee J H and Kim Y M 2003 *Carbohydr. Res.* **338** 1183.
[8] Capek P, Hlavonova E, Matulova M, Mislovicova D, Ruzicka J and Koutny M 2011 *Carbohydr. Polym.* **83** 88.
[9] Shingel K I 2002 *Carbohydr. Res.* **337** 1445.

IOP Conf. Series: Materials Science and Engineering 1280 (2023) 012010 doi:10.1088/1757-899X/1280/1/012010

Preparation and Characterization of Novel Membrane from Waste Polyethylene terephthalate and Bio-based Polymer

L Jareonsri[1,*], S Nawalertpanya[1] and W Jantaporn[1]

[1] Department of Chemical Engineering, King Mongkut's University of Technology Thonburi, 126 Pracha Uthit Rd., Bang Mod, Thung Khru, Bangkok 10140, Thailand.

*Corresponding author: lapasrada.jar@mail.kmutt.ac.th

Abstract. A large amount of post-consumer Polyethylene terephthalate (PET) bottle waste could be recycled for downstream products. In this work, used PET bottles were utilized as a raw material to develop environmentally-friendly ultrafiltration membranes by phase inversion technique. Poly(ethylene-2,5-furandicarboxylate) (PEF), prepared from a sustainable bio-based monomer, 2,5-furandicarboxylic acid (FDCA), was blended into rPET matrix to improve the mechanical properties of the membrane. Ethanol was also used as a non-solvent in a coagulation bath and polyethylene glycol (PEG) with a molecular weight of 400 Da was added as a hydrophilic additive. The effects of PEF and PEG incorporation to waste PET-based membranes have been thoroughly studied. The morphology of obtained membrane was investigated using Scanning Electron Microscopy (SEM). The membrane's permeation and resistance were tested using a pressure-driven dead-end membrane module. The result revealed that adding PEF increased the roughness of the membrane surface, which lowered the flux recovery ratio and rejection performance. However, the filtration performance and the membrane resistance could be improved by adjusting the content of the PEG additive.

1. Introduction

The consumption of post-consumer Polyethylene terephthalate (PET) has potentially risen today and tends to increase significantly. According to an analysis conducted by the Plastics Institute of Thailand, only 31% of PET bottle is collected for recycling and 63.4% of the PET recycling value is lost by producing lower-grade products [1]. Thus, the need to increase the value of recycled PET (rPET) is still in high demand. Recent studies showed the possibility of membrane fabrication from rPET by phase inversion process. The success of developing rPET membranes would be a green alternative to membrane technology. Because the commercial polymer materials used for membrane production today such as polysulfone (PSF), polyethersulfone (PES), and polyvinylidene fluoride (PVDF) are fossil-based polymers which lead to high manufacturing costs and a future fossil depletion crisis.

The results from recent publications showed that membranes prepared from recycled PET bottles required polymeric additives. Without additives, the obtained rPET membranes possess high rigidity, weak mechanical, and lack of hydrophilicity properties which are essential for further membrane utilization [2-4]. This work aims to create novel sustainable attractive membranes from zero carbon footprint sources. Therefore, combining bio-derived polymer with waste fossil-based polymer, rPET, is one of the ways to reach environmental sustainability and reduce the consumption cost of fossil-based polymer. Recently, FDCA has become a promising bio-based compound. PEF prepared from FDCA and ethylene glycol is a new polymer with comparable properties to PET. The studies on the copolymerization of PET and PEF monomers showed improvements in thermal and mechanical properties [5-10]. However, the study about this copolymer application is still limited. To the top of our knowledge, there is still no PET/PEF membrane preparation publication. In this work, PET/PEF membrane casting solutions are prepared using polyethylene glycol (PEG) with a molecular weight of 400 g/mol as a plasticizer additive. The effect of casting solution concentration on membrane microstructure, hydrophilicity, and performance are thoroughly investigated.

2. Experimental

2.1 *Materials*

PET was obtained from clear waste plastic bottles cut into flakes by Prabkaya Recycle, Pathum Thani Thailand. Poly(ethylene-2,5-furandicarboxylate) (PEF) was synthesized from 2,5-furandicarboxylic acid (2,5-FDCA, 99%) purchased from Hebei Chisure Biotechnology and ethylene glycol (EG, analytical reagent grade) as previously reported [8]. Phenol (≥ 99%) was used as the solvent and supplied by Loba Chemie. Polyethylene glycol with a molecular weight of 400 Da (PEG400) was used as the additive and obtained from Sigma-Aldrich. Ethyl alcohol (analytical reagent grade, ≥ 99%) was used as the non-solvent. In the rejection experiment, the Albumin Bovine (pH 6-7) fraction V (Bovine Serum Albumin, BSA), 98% was used and supplied by Srichem.

2.2 Casting solution preparation

First, waste PET bottle flakes were washed with water several times and dried at 70°C using a hot air oven for 48 h. Phenol was liquified by heating at 40°C. Then, the dried PET flakes were dissolved in phenol at 100°C under a continuous stirring using a magnetic bar. PEF and the additive, Polyethylene glycol (PEG), were separately dissolved in phenol at the same condition. After obtaining homogeneous solutions, the solutions were mixed and heated at 100°C under continuous stirring for one hour. The casting solution was then heated without stirring for 15 min. The compositions of casting solutions are listed in Table 1. The membranes tested were assigned as M "X"-"Y", where X and Y corresponded to their PEF and PEG weight ratio contents, respectively.

Table 1. The composition of membrane casting solution.

Membrane Name	Weight ratio				Concentration (wt%)	
	PET	PEF	PEG400	Phenol	PEF	PEG400
M 0-2	4	0	2	20	0	7.69
M 1-2	4	1	2	20	3.70	7.41
M 2-2	4	2	2	20	7.14	7.14
M 1-1.5	4	1	1.5	20	3.77	5.66
M 1-1	4	1	1	20	3.85	3.85

2.3 rPET-PEF Membrane preparation

The freshly prepared casting solution was immediately cast onto a clean glass plate using a glass casting rod. The glass plate with the polymeric film was immediately immersed in a coagulation bath (containing ethanol as a non-solvent). After 20 min, membranes were taken out and rinsed several times with deionized water. Membranes were kept in containers containing deionized water for further use. The membranes were visually inspected before use to ensure no defective areas in the membranes.

2.4 rPET-PEF Membrane characterization

Scanning electron microscope (SEM, JSM-6610 LV) was used to characterize the membrane's morphology. The thickness of the membranes was measured using a digital thickness gauge (Keiba). The reported values in section 3 were the average value of five different positions.

2.5 Filtration experiment
The filtration experiment was conducted using a pressure-driven dead-end module utilizing the pressure from nitrogen gas as a driving force. The operating condition was set at 1 bar. In the first part, deionized water was used as a feed to study pure water flux (PWF). The second part was the study of the membrane's ability to reject BSA. Deionized water with a BSA concentration of 1g/L was used, and the rejection (R) was determined. The sample of permeate was taken when the steady-state condition was achieved. A UV-Vis spectrophotometer (Hitachi, U-3000) was used to measure the BSA concentrations. The membranes fouling resistance were also characterized by determining the flux recovery ratio (FRR), flux decline ratio (R_t), reversible fouling (R_r), and irreversible fouling (R_I). The equations were followed from Abdel-Karim's work [11].

3. Results and discussion

3.1 Effect of the PEF % content as sub-polymer
Firstly, the effect of PEF sub-polymer % content varied from the weight ratio of 0,1 and 2. As pictured in Figure 1, the SEM images showed that the membrane had more porous structure when PEF % content increased from M.1-2 to M.2-2. The similarity was also found in other works where increasing sub-polymer concentration enlarges the pores in the support layer [12-14]. Nonetheless, the membranes were still dense and smooth on the surface (or the active layer). Except for M.2-2, the high roughness morphology could be observed. Because adding hydrophilic additives to casting solution more than a specific value (3.70% in this study) slows down the exchange rate between solvent and non-solvent. As a result, the surface chemistry parameters are affected [15]. Thus, the high surface roughness on M.2-2 was observed. This high surface roughness could result in the decline of permeate flux.

Moreover, the membrane thickness was increased with the addition of the PEF sub-polymer. A similar result was also reported by Kiani et al. [12]. Solidification during the membrane formation could affect to the membrane thickness. Since, the chain length of the polymer molecule is larger and the viscosity of the solution is increased while increasing the solid content. This reduces molecular mobility, and consequently higher membrane thickness was obtained [12, 15, 16].

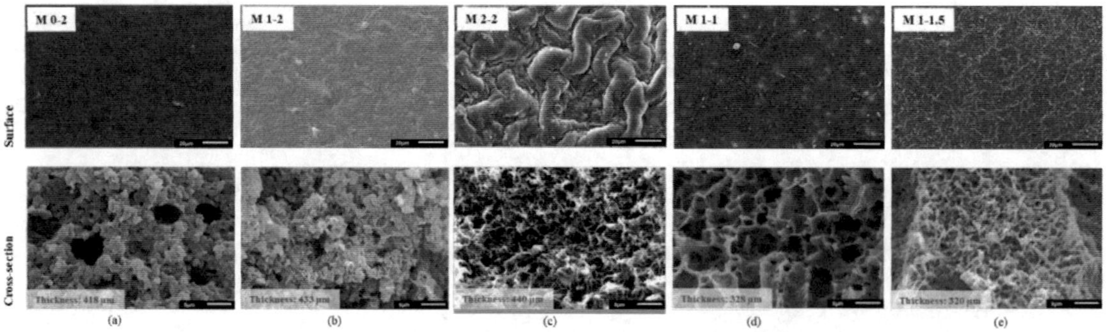

Figure 1. SEM images of rPET/PEF/PEG membranes at different concentrations.

Figure 2a represents the water permeate flux (0-180 min), along with the permeate flux from aqueous BSA solution (181-360 min), followed by permeate flux of DI water (361-390 min). The membrane resistance and rejection are shown in Figure 2b. The water permeate flux of the membrane increased with the increase in the concentration of the PEF sub-polymer. This result is in accordance with the microstructure analysis from SEM images as described above. However, the roughness of the M.2-2 membrane surface resulted in decline of permeate flux and the fouling of BSA which accumulated on the membrane surface. The decline was clearly observed in M.2-2. The increase in PEF content resulted in a decrease in R_r values as well as FRR. On the other hand, the total fouling of membranes had no significant difference.

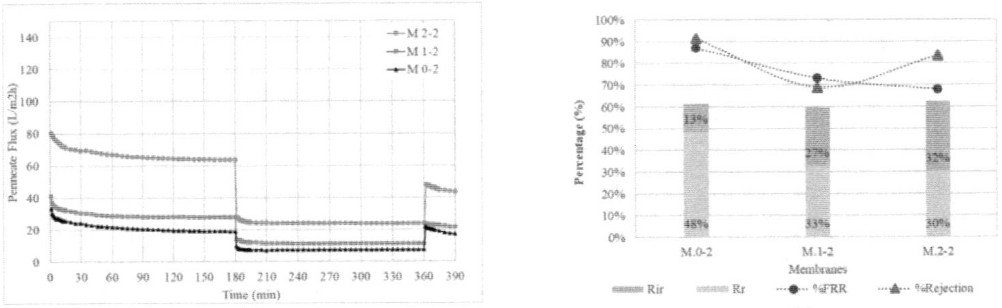

(a) (b)

Figure 2. Permeate fluxes (a), and resistant (b) of membranes with different PEF concentrations.

3.2 Effect of the PEG % content as additive

The composition between PET/PEF of 4/1 was chosen for further study. Due to the most outstanding morphology and higher FRR value than M.2-2. The weight ratio of PEG varied from 1, 1.5, to 2. As can be seen in Figure 1, the porosity in the cross-section was increased with the addition of PEG concentration. The addition of PEG concentration (from M.1-1 to M.1-1.5) led to the decreasing of membrane thickness. Since the thickness of the membrane depends on its structure and morphology [16, 17]. The addition of a high hydrophilic additive helps slow down the membrane formation in the coagulation bath. Still, increasing hydrophilic additive higher than a specific value led to accelerating the membrane formation process [18]. Thus, an increase in membrane thickness was observed in M 1-2.

From Figure 3a, the permeate flux surprisingly decrease when decreasing the PEG400 concentration to the weight ratio of 1 (M.1-1). However, the water permeate fluxes of M.1-1 and M.1-1.5 were similar. The BSA rejection values and the membrane-resistant values are shown in Figure 3b. As can be seen, reducing the concentration of PEG400 from the weight ratio of 2 to 1 increased BSA rejection. With 81% of flux recovery, M.1-1.5 showed the lowest in total fouling at 45%, where 26% of them was R_r. This indicated that the membrane fouling could be escaped by simple hydraulic cleaning. However, even M.1-1.5 and M.1-2 resulted in similar rejection values (71% and 69%). The flux recovery ratio and the total fouling were very different. The flux recovery of M.1-1 is lowest at 49%. This indicated that M.1-1 had the lowest anti-fouling property compared to others. Even M.1-2 showed the highest flux with a 73% FRR value, most of the fouling was irreversible which required special cleaning methods such as chemical or enzymatic cleaning. Thus, by comparing these three rPET/PEF/PEG compositions, M.1-1.5 showed the best improvement of anti-fouling property even though the flux was extremely declined from M.1-2.

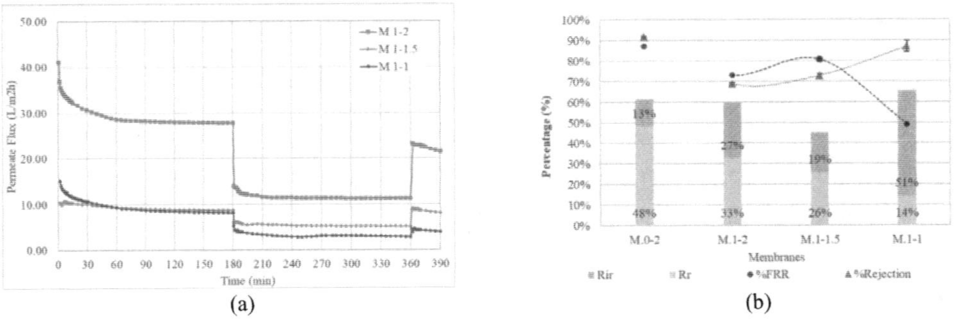

(a) (b)

Figure 3. Permeate fluxes (a), and resistant (b) of membranes with different PEG400 concentrations.

4. Conclusion

The novel membrane from recycled polyethylene terephthalate (rPET) and bio-based polymer, poly-(ethylene-2,5-furandicarboxylate) (PEF), was successfully prepared. The addition of PEF content into the PET membrane resulted in the increasing of water and BSA permeate flux. Without adjusting the PEG additive content, the prepared membranes showed a decrease in flux recovery ratio and rejection values. By observing the effect of PEG concentrations, the membrane prepared from rPET/PEF/PEG of 4/1/1.5 showed the best improvement in the anti-fouling property. However, increasing the % rejection as well as reducing irreversible fouling value should be considered in further study. Thus, the results from this study show that combining bio-derived polymer with waste fossil-based polymer should be developed to have great potential for the industry sector in the future.

Acknowledgment

The authors would like to thank the Department of Chemical Engineering at King Mongkut's University of Technology Thonburi for the full scholarship on tuition fees in Academic-Research Chemical Engineering Practice School (AR-ChEPS) program.

References

[1] W.B. Group, Market Study for Thailand : Plastics Circularity Opportunities and Barriers, East Asia and Pacific Region Marine Plastics Series, Washington, DC 2021.
[2] Rajesh S and Murthy, Z V P 2014 *Quím. Nova.* **37** 653.
[3] Kusumocahyo S P, Ambani S K, Kusumadewi S, Sutanto H, Widiputri D I and Kartawiria I S 2020. *J. Environ. Chem. Eng.* **8** 104381.
[4] Kusumocahyo S P, Ambani S K and Marceline S 2021 *Sustain. Environ. Res.* **31** 19.
[5] Sun L, Zhang Y , Wang J, Liu F, Jia Z, Liu X and Zhu J 2019 *J. Appl. Polym. Sci.* **136** 47186.
[6] Sousa A F, Vilela C, Matos M, Freire C S, Silvestre A J and Coelho J F 2015 *Compos. Nanocompos*. 113.
[7] Wang P, Huang W, Zhang Y, Lin J and Chen P 2020 *J. Polym. Sci.* **58** 320.
[8] Papadopoulos L, Zamboulis A, Kasmi N, Wahbi M, Nannou C, Lambropoulou D A and Bikiaris D N 2021 *Green Chem.* **23** 2507.
[9] Sousa A F, Matos M, Freire C S, Silvestre A J and Coelho J F 2013 *Polymer* **54** 513.
[10] Kasmi N, Majdoub M, Papageorgiou G Z, Achilias D S and Bikiaris D N 2017 *Polymers.* **9** 607.
[11] Abdel-Karim A, Leaper S, Alberto M, Vijayaraghavan A, Fan X, Holmes S M and Gorgojo P 2018 *Chem. Eng. J.* **334** 789.
[12] Kiani S, Mousavi S M, and Bidaki A 2021 *J. Clean. Prod.* **314** 128082.
[13] Rajesh S and Murthy Z V P 2015 Desalination and Water Treatment **57** 1-10.
[14] Sun Z and Chen F 2016 *Int. J. Biol.* Macromol. **91** 143.
[15] Kiani S, Mousavi S M, Saljoughi E and Shahtahmassebi N 2018 *Polym. Adv. Technol.* **29** 1632.
[16] Saljoughi E, Sadrzadeh M and Mohammadi T 2009 J. Membr. Sci. **326** 627.
[17] Dan Y, You Y, Wei L and Han T 2011 *Chinese J. Polym. Sci.* 26.
[18] Jullok N, Darvishmanesh S, Luis P and Van der Bruggen B 2011 *Chem. Eng. J.* **175** 306.

Theoretical Analysis of High-Harmonic Generation in Silicon Nanotubes

C Summueang[1], O Suntijitrungruang[1] and S Boonchui[1,2,*]

[1]Department of Physics, Faculty of Science, Kasetsart University, Bangkok 10900, Thailand 10900
[2]Center of Rubber and Polymer Materials in Agriculture and Industry (RPM), Faculty of Science Kasetsart University, Thailand

*Corresponding author: fscistb@ku.ac.th

Abstract. Silicene hexagonal nanotube (Si h-NT) is a one-dimensional periodic system consisting of the rolling silicene layer, a monolayer of silicon atoms. In this research, we explore high-order harmonic generation (HHG) with a strong mid-infrared (IR) field on the single-walled nanotube. Moreover, the electronic intraband and interband dynamics are significantly and especially investigated in order to study the HHG mechanisms thoroughly. We then show and discuss the numerical results of the HHG characteristics from Si h-NT compared with single-walled carbon nanotubes and silicene sheets.

1. Introduction

Silicene, the silicon-based counterpart of graphene, possesses a two-dimensional structure that has a variety of potentially applicable physical and chemical attributes [1]. To grow a silicene sheet, it can employ scanning tunnelling microscopy for studying experimentally the evidence of a silicene sheet [2]. In theoretical aspects, the density-functional based tight-binding theory is used for analyzing the properties of silicon hexagonal [3] such as the mechanical, electronic, energetic, and structural characteristics. Significantly, the work [4] portrays that the tight-binding theory is a useful framework to investigate the electronic properties of silicene. Similar to graphene [5,6] which can be synthesized to be nanotubes with different chiral types [8,9], so also silicene [6]. Zigzag silicon hexagonal nanotube (Si h-NT) [10,11] is the chiral type that the research [12] studied theoretically through the tight-binding method. Moreover, silicene certainly can generate high harmonic photon emission [14,16] in common with graphene and carbon nanotubes [7]. This phenomenon was considered by specific approaches like the semi-classical calculations that analyze the electron transition (both from interband and intraband components) [13,15] and the time-dependent Schrödinger equation from the tight-binding method and dipole effect. Based on these former appropriate methods in this work, we employ such a framework to explore theoretically a high harmonic generation in zigzag silicene nanotubes.

2. Theoretical method

Single-walled Si h-NT is provided by rolling up a monolayer silicene sheet just as a carbon nanotube constructed by rolling up a graphene. The nanotube refers to a silicene layer rolled along the direction identified by the chiral vector $\mathbf{C}_h = n_1 a_1 + n_2 a_2$ where a_1 and a_2 are primitive hexagonal lattice vectors with integers n_1 and n_2 showing in Fig.1(a). The chiral indices (n_1, n_2) determine the nanotube characteristic. The armchair type corresponds to the chiral indices (n, n), while (n,0) denotes the zigzag type. In addition, the nanotubes are defined in the coordinates (x, y), where the circumference is considered in the x direction: $k_\perp$ (parallel to chiral vector $\mathbf{C}_h$) and the nanotube axis in y direction: $k_\parallel$ (perpendicular to chiral vector $\mathbf{C}_h$). A nanotube unit cell as a cylindrical segment of length T is a translation vector in the y-direction. The wave vector in the y-axis within the first Brillouin zone is $-\pi/T \le k_\parallel \le \pi/T$. The periodic boundary condition on the wave vector $k_\perp$ discrete is defined as $k_\perp = 2\pi l/|\mathbf{C}_h|$ in different band indices l which can be implied in the following Ref. [1,4,11].

Figure 1 (a) Schematic plan and (b) top view of rolled up of the silicene monolayer.

In our model, we focus on the zigzag Si h-NTs corresponding to a pair of indices (n, m = 0). We apply the TB models to obtain the wave functions and band structures of single-walled Si h-NTs. The Hamiltonian H_0 can be explained by the 2NN interactions following [7]. The time-dependent Hamiltonian $H(t) = H_0 + V_i(t)$ where $V_i(t) = -q_e F(t)\,\hat{y}$ is the driving electric field. Therefore, representing + (−) as the conduction (valence) band, the wave function can be expressed as

$$|\Psi(t)\rangle = \int \Psi_k(t)\,dk = \int\left[C_+(\mathrm{k,t})\left|\Phi_{\mathrm{k}}^+\right\rangle + C_-(\mathrm{k,t})\left|\Phi_{\mathrm{k}}^+\right\rangle\right]dk \tag{1}$$

Inserting Eq. (1) into TDSE: $i\hbar\frac{\partial}{\partial t}|\Psi(t)\rangle = H(t)|\Psi(t)\rangle$ and the time-dependent momentum can be written as $\boldsymbol{\kappa}_t = k - \left(q_e/\hbar\right)A(t)$ where $A(t)$ is vector potential. The total electric field is determined as $F(t) = F_0 \sin^2\left(\pi t/8T\right)\sin(\omega_0 t)$ with $0 \le t \le 8T$, F_0 is field amplitude, T is the period and ω_0 is field frequency. We derive the set of two differential equations in the form:

$$\begin{aligned} i\hbar\frac{d}{dt}C_+(\boldsymbol{\kappa}_t,t) &= \left[E_+(\boldsymbol{\kappa}_t) - F(t)\cdot D(\boldsymbol{\kappa}_t)\right]C_+(\boldsymbol{\kappa}_t,t) - F(t)\cdot D(\boldsymbol{\kappa}_t)C_-(\boldsymbol{\kappa}_t,t) \\ i\hbar\frac{d}{dt}C_-(\boldsymbol{\kappa}_t,t) &= \left[E_-(\boldsymbol{\kappa}_t) - F(t)\cdot D(\boldsymbol{\kappa}_t)\right]C_-(\boldsymbol{\kappa}_t,t) - F(t)\cdot D(\boldsymbol{\kappa}_t)C_+(\boldsymbol{\kappa}_t,t) \end{aligned} \tag{2}$$

where $D(\boldsymbol{\kappa}_t)$ is the transition dipole matrix element between conduction and valence band in Ref. [15]. The expectation value of dipole consists of two transitions by taking the second derivative to solve the harmonic emission by substituting the wave function that can be written in the following form:

$$d(\boldsymbol{\kappa}_t) = \langle\Psi(\mathrm{t})|q_e\hat{\mathbf{r}}|\Psi(\mathrm{t})\rangle = \int\left[d_\updownarrow(\boldsymbol{\kappa}_t,t) + d_\leftrightarrow(\boldsymbol{\kappa}_t,t)\right]dk \tag{3}$$

where $d_\updownarrow(\boldsymbol{\kappa}_t,t)$ is interband transition and $d_\leftrightarrow(\boldsymbol{\kappa}_t,t)$ intraband transition as [15] and the probability coefficients need to satisfy the normalization condition $|C_+(\boldsymbol{\kappa}_t,t)|^2 + |C_-(\boldsymbol{\kappa}_t,t)|^2 = 1$.

3. Results and discussion

The band structure of zigzag Si h-NT is shown in Fig.2(a). This simulation is the eigenvalue of Hamiltonian. From these graphs, it is noticeable that the energy of zigzag Si h-NT is not symmetrical along the y-axis despite the same band index. Such an asymmetrical characteristic is due to the chiral effect of rolling nanotubes. To investigate HHG, Fig.2(b) demonstrates the harmonic spectrum from the electron transition in the band indices $l = 1$ and $l = 2$. Generally, the transition can be categorized into two different kinds: the interband and intraband transition. From Fig.2(b), it illustrates that the HHG from the intraband transition is more intense than HHG of the interband transition. This tendency can be described to the different probability of the electron's transition. For these band indices, the interband emission is harder than the intraband emission.

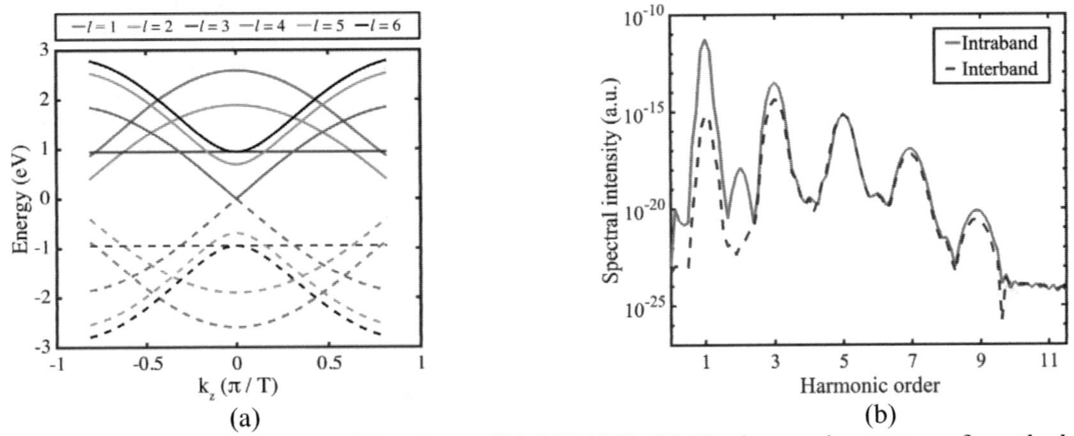

Figure 2 (a) The band structure of the zigzag Si h-NT (6,0). (b) The harmonic spectrum from the band index $(l = 1)$ by a $3\mu m$ wavelength with the driven field intensity $5\times10^{10}\ W / cm^2$.

Considering the band index $l = 3$, this is a flat band according to Fig.2(a). As a result, the energy released from intraband transition should be much less than the energy released from interband transition. As illustrated in Fig.3(a) that is the harmonic spectrum from the band index $l = 3$, it is shown vividly that the HHG of the intraband emission is less than the interband emission. Thus, this result conforms to the energy band structure.

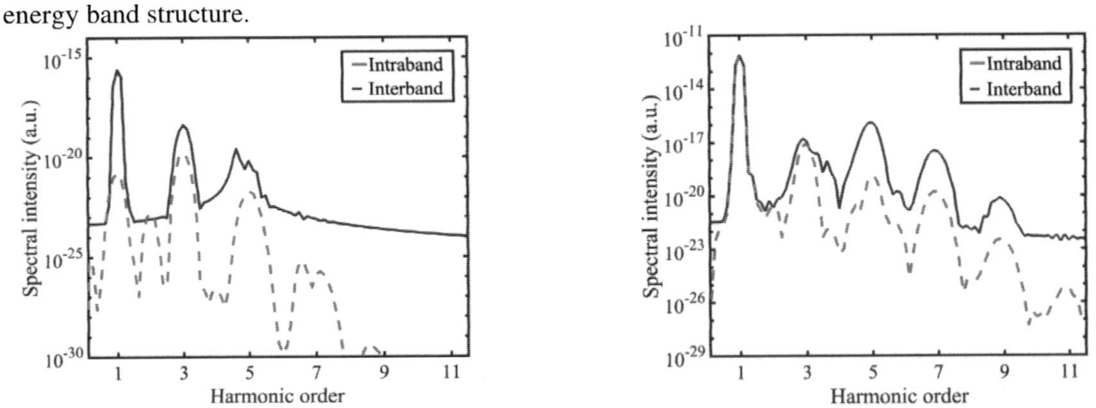

Figure 3 The harmonic spectrum from (a) band index $l = 3$ and (b) band index $l = 5$.

Finally, we consider the HHG from the band indices $l = 5$ and $l = 6$ as shown in Fig. 3(b). In this case, HHG from the interband transition is more intense than the intraband transition. From Fig.2(a), the different energy (between conduction band and valence band) in these band indices is less than the different energy in the band indices $l = 1$ and $l = 2$. For this reason, the probability of interband emission from the band indices $l = 5$ and $l = 6$ is more related to the probability from the band indices $l = 1$ and $l = 2$, leading to more HHG from the interband transition. In addition, the HHG results in the band indices $l = 5$ and $l = 6$ is similar to the HHG from CNT in Ref. [15]. Furthermore, the HHG from silicene nanotubes (in every band index) originates from odd harmonic numbers significantly. As a result of the symmetry property of the Brillouin zone in silicene nanotubes, the energy released from even harmonic frequency is very little as demonstrated in our simulation.

4. Conclusions

By the Schrodinger equation with a tight-bind model, we derive the harmonic spectrum for silicene nanotubes. From the theoretical calculation in our method, chiral and band indexes of nanotubes impact on the characteristics of the energy band structure. We, focussing on the zigzag chiral type, notice that the energy band structure of this chiral type is asymmetric owing to chiral effect. Concerning the HHG, the photon emission stems from two different kinds of electron's transition that are intraband transition and interband transition. For certain band indexes such as $l = 1$ and $l = 2$, the most HHG comes from the intraband transition since the interband transition, in these band indexes, is harder than the intraband transition due to high different energy between conduction band and valence band. On the other hand, for particular band indexes (like $l = 5$ and $l =6$) where the difference energy between conduction band and valence band is lower, the most HHG originates from the interband transition because the interband transition in these band indexes takes place more easily. In addition, for the band index $l = 3$ where the energy band is flat, the HHG comes from interband transition significantly since the energy released from intraband transition is very low compared to the energy released from the interband transition. The HHG in silicene nanotubes, besides, originates from odd harmonic numbers mostly. Due to the symmetry property of the Brillouin zone in silicene nanotubes, the emission of even harmonic numbers vanishes. From our work, we demonstrate that the tight-binding framework is useful and plausible to analyze the HHG in silicene nanotubes. Our research can be applied to the astrophysics and high frequency lasers for various benefits.

Acknowledgement

This work was supported by Science Achievement Scholarship of Thailand (SAST), specialized center of rubber and polymer materials in agriculture and industry (RPM) and Faculty of Science Kasetsart University for partial support.

References

[1] Kara A, *et al.*, 2012 *Surf. Sci. Rep.* **67** 1.
[2] Lalmi B, *et al.*, 2010 *Appl. Phys. Lett.* **97** 223109.
[3] Seifert G, *et al.*, 2001 *Phys. Rev. B* **63** 193409.
[4] Guzmán-Verri GG and Voon LCLY 2007 *Phys. Rev. B* **76** 075131.
[5] Suntijitrungruang O, Summueang C and Boonchui S 2021 *Opt. Mater. Express* **11** 2854
[6] Fanbanrai P, Hutem A and Boonchui S 2015 *Physica B: Condens. Matter* **472** 84.
[7] Charoenpakdee J, Suntijitrungruang O and Boonchui S 2020 *Sci Rep* **10** 18949.
[8] Thitapura T, Liewrian W, Jutarosaga T and Boonchui S 2017 *Opt. Express* **25** 25588.
[9] Thitapura T, Liewrian W, Jutarosaga T and Boonchui S 2017 *Plasmonics* **12** 1439.
[10] Ahmadi N, Shokri A A, Elahi S M 2016 *Silicon* **8** 217.
[11] Ezawa M 2012 *EPL. Europhys. Lett.* **98** 67001.
[12] Shokri A and Ahmadi N 2015 *Opt. Quant. Electron* **47** 2169.
[13] Vampa G, *et al.*, 2014 *Phys. Rev. Lett.* **113** 073901.
[14] Qin R and Chen Z Y 2018 *Nanoscale* **10** 22593.
[15] Zurrón-Cifuentes Ó, *et al.*, 2020 *Opt. Express* **28** 19760.
[16] Summueang C, Fudulwatjananon T and Boonchui S 2022 *IOP Conf. Ser.: Mater. Sci. Eng.* **1234** 01203.

IOP Conf. Series: Materials Science and Engineering 1280 (2023) 012012 doi:10.1088/1757-899X/1280/1/012012

Synthesis of Self-organic Assembly Monolayers to Improve the Stability of Perovskite Films

P Phonglumjiakngam and N Pootrakulchote*

Department of Chemical Technology, Faculty of Science,Chulalongkorn University, Bangkok 10330, Thailand

*Corresponding author: nuttapol.p@chula.ac.th

Abstract. In recent years, photovoltaic devices based on perovskite materials have become one of the energy conversion field's most in study topics. $FAPbI_3$ stands out among a wide variety of perovskite structural materials with the advantages of a suitable band structure, long carrier diffusion distance, and so on. Unfortunately, the $FAPbI_3$ perovskite films still lack some stability. It is well known that UV light and humidity strongly affect perovskite solar cell (PSC) performances, leading to the service life-time limitation of the device. Here, we propose strategies to improve the stability of PSCs by means of the interface engineering [1] between the electron transport layer and the perovskite active layer by inserting the self-organic assembled monolayers (SAMs). [2] The derivatives of Triethoxy(4-methoxyphenyl)silane (SAM1) and Triethoxy(3,4-dimethoxyphenyl)silane (SAM2) were synthesized using Grignard reaction [3] then inserted between the electron transport layer and perovskite active layer of PSC test devices in order to improve the hydrophobicity of the cell. Preliminary results showed that the perovskite active layers that have been modified with SAMs appear denser than those without modification. The overall conversion efficiency and stability of the SAMs-modified PSC devices will be further tested to prove more stable than those of unmodified devices.

1. Introduction

Nowadays, Perovskite solar cells (PSCs) is a promising candidate technology for the next generation of solar cells due to their high charge populations, narrow band gap, high charge mobility and low excitation energy [4]. Generally, High-performance perovskite solar cells are typically built with layers of fluorene doped tin oxide (FTO) coated glass, Electron Transport Layer (ETL) , perovskite, Hole Transport layer (HTL), and metal contacts, following the trend of published papers. In the last ten years, perovskite solar cell efficiency has increased dramatically, from 3.8% to 24%. [6] But the current focus of most perovskite solar cell research is still on improving photoconversion efficiency (PCE). However, the stability issue is one important problem that keeps PSCs away from becoming commercially successful. In perovskite solar cells, the ETL plays critical roles in both harvest and transportation of photogenerated electrons, as well as the inhibition of electron-hole recombination. So, The ideal of electron transport layer should have an good optical transparency, high performance in transporting electrons and blocking holes, good electrical conductivity, and a well-matched energy band with perovskite materials. [5] Perovskite plays the role of an active layer that absorbs light and generates

IOP Conf. Series: Materials Science and Engineering 1280 (2023) 012012 doi:10.1088/1757-899X/1280/1/012012

electrons in the cell. The direct contact of these two layers causes a problem, which makes perovskite film decompose into the PbI2 precursor after the film is thermally annealed. [2] Here, in this study, we introduce the use of Self-organic Assembled Monolayer (SAM) as an interfacial layer inserted between ETL and perovskite active layer to improve the stability of perovskite solar cell.

2. Experimental

2.1 Preparation of ETL

In this work, titanium dioxide is used as an ETL. The preparation of titanium dioxide was successfully done by adding TiO_2 commercial (Titanium (IV) oxide, anatase, nano powder, <25 nm particle size, Sigma-Aldrich) to the binder solution, which was prepared in the ratio of 1:37.5 by weight of alpha-terpineol and Ethyl cellulose in the ratio of 1:4 by weight.

2.2 Synthesis of SAMs

SAM1 and SAM2 were synthesized using Grignard reaction incorporated with 4-Bromomagnesium Anisole and 4-Bromomagnesium Veratrole, respectively. Frist, a Magnesium turning (134 mg, 5.50 mmol) and Dried THF (1.0 mL) was added to glass bottom. Next, the galss bottom was charged with 4-bromoanisole (5.00 mmol) or 4-Bromoveratrole (5.00 mmol) in 2.0 mL of THF. With a condition that reaction was initiated by adding 5-10 drops of the bromoanisole or bromoveratrole solution to the magnesium turnings with stirring, followed by gentle heating. Then the rest of the bromoanisole or bromoveratrole solution was then added at such a rate that the THF maintained a moderate reflux. The solution was stir at -30oC for 1 h and then at room temperature for 12 h. The crude reaction mixture was then poured into 50 mL of pentane in a 200 mL separatory funnel. The yellow solution was washed with 3 × 25 mL of DI water, dried over MgSO4, filtered, and the residue was purified by distillation at 200-275 0C in vacuum condition. SAMs were characterized using 400 MHz BRUKER NMR, Thermo Fisher FT-IR Spectrometer and Thermo Fisher Scientific UV-Vis Spectrophotometer.

2.3 Preparation of a Perovskite Precursor Solution

The absorber layer (Perovskite) was deposited from a precursor solution containing lead (II) iodide 0.461 g, Formamidinium 0.172 g and Cesium chloride 0.033 mg in mixed solvent of DMF and DMSO, of which the volume ratio is 1:4. The solution was then stirred at 60°C for 15 minutes.

2.4 Fabrication of Perovskite Films

To prepare the full cell, FTO coated glass (size 3×3 cm, thickness 3 mm) was cleaned with detergent and rinsed with deionized water. After that, the substrates were washed with Isopropanol and Ethanol in an ultrasonic bath for 30 minutes. then dried using UV-Ozone. The solution of ETL was preapred by 1 ml of TiO_2 (from section 2.1) stirred with 5 ml of ethanol for 20 h. After that, 400 μL of ETL solution was spin-coated at 2,100-2,500 rpm for 30 seconds on top of FTO glass and heated at 500°C for 30 min under an ambient condition. SAM1 or SAM2 was coated on the ETL using dipping method. The FTO glass/TiO_2 substrate was soaked in the SAMs solution for 24 h. and washed with methanol to remove the physically adsorbed SAM molecules. After washing, the substrate dried under 50°C heated under ambient air.

The perovskite precursor solution was spin-coated on FTO glass/TiO2 substrate by two step method of spin-coated program at 1,000 rpm and 6,000 rpm for 20 and 30 respectively. During the second step, 100 μL of chlorobenzene was gentle dropped onto the spinning substrate 10 s after the start of program. Finally, the film was annealed on the hotplate at 185°C for 20 min.

2.5 Characterization of SAM-modified perovskite

The characterization of SAM-modified perovskite was done using the water contact angle, SEM images, and UV-Vis spectra.

3. Result and Discussion

3.1 Synthesis and Characterization of SAMs

The synthesis of SAMs (derivatives of aryl siloxanes) has been accomplished by the treatment of Grignard reagents with tetraethyl orthosilicate. The product yield of the synthesis was 15–25% as a colorless oil. The NMR spectra in Figure 1 show that the SAM2's spectrum matches that of SAM1, but there appears an additional peak (as shown in the dashed circle) that is assigned to the other methoxy group that has been formed during the reaction with some unknown impurities.

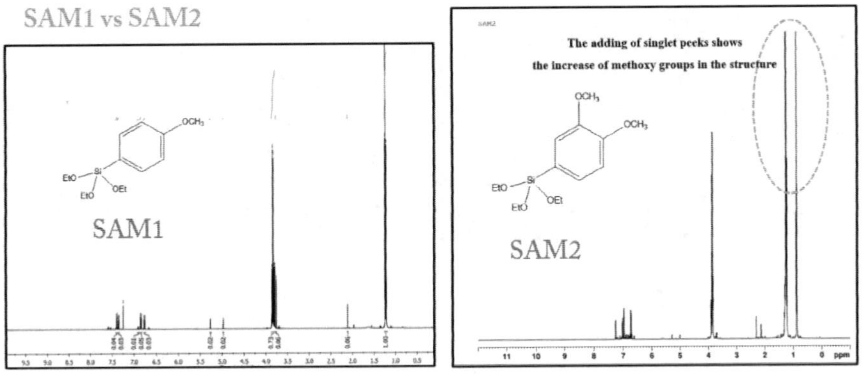

Figure 1. The NMR spectra of Synthesized SAM1 and SAM1

3.2 Characterization of Cell Fabrication

SAMs are modified onto TiO_2 by dipping and then by the silanization reaction, which uses methanol as a solvent to convert the oxygen atoms of TiO_2 and O-ET-Si of SAMs to hydroxy groups (-OH) and react with each other to form the Ti-O-Si bond between SAMs and TiO_2.

Perovskite typically reacts strongly with moisture, causing degradation. To further confirm the deposition of SAMs on the surface of TiO_2, the surface wettability of the unmodified and modified TiO_2 ETLs was investigated by measuring the water contact angles on each of TiO_2 ETL surface.

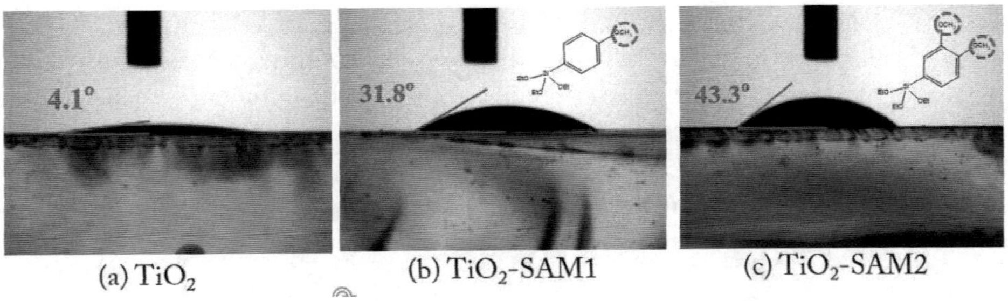

Figure 2. Water contact Angle of Modified SAMs on TiO_2 ETL surface

As shown in Figure 2, the water contact angel on TiO_2 ETL surface significantly increased from 4.1 to 31.8 and 43.3 degrees after being modified with SAM1 and SAM2, respectively, suggesting the enhanced hydrophobic properties of TiO_2 surfaces. This is probably because the hydrophilic hydroxyl groups on the TiO_2 surface were replaced by the more hydrophobic methoxy groups of SAMs. The effect of SAMs treatment on the morphology of the perovskite films coated on the TiO_2 ETLs were studied using top-view SEM images. As shown in Figure 3., larger perovskite grain boundaries were observed in case of SAMs modified TiO_2 compared to those of the pristine TiO_2. This is attributed to the reduction of the TiO_2 surface wettability. Furthermore, the density of perovskite apparently increases as a result of an increasing number of hydrogen bonds between the methoxy groups in SAMs and the formamidinium (FA) cations in perovskite, effectively suppressing the deprotonation reaction between the formamidinium cation and the hydroxyl group of TiO_2, which leads to perovskite material degradation.[2]

Figure 3. SEM image results that show the effect of SAMs modified on TiO_2 ELT

3.3 The Effect of SAM Modification on Perovskite Active Layer Stability

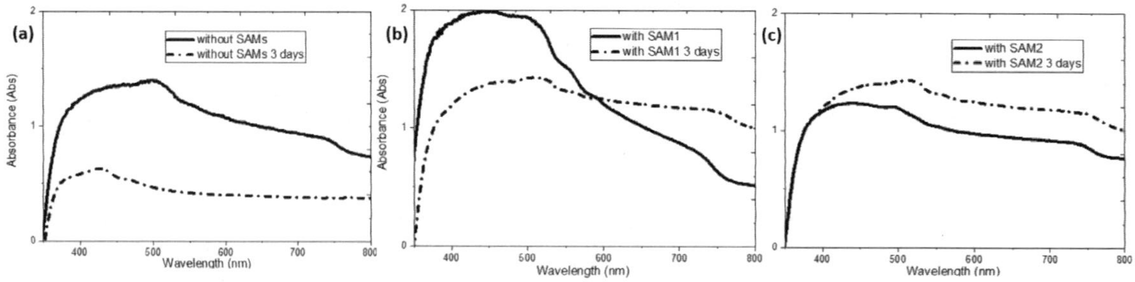

Figure 4. UV-Vis Absorption Spectra of SAMs on Perovskite: (a) Perovskite unmodified, (b) Perovskite with SAM1 modified and (c) Perovskite with SAM2 modified

The UV-Vis absorption spectra of the perovskite films with and without SAMs modification is shown in Figure 4. As seen from Figure 4a., the absorption ability of the unmodified perovskite film drops remarkably after three days in the dark. In comparison, the absorption of the perovskite films that have been modified with SAM1 and SAM2 after three days kept in the dark did not change significantly as shown in Figure 4b. and 4c. The preliminary results suggest that SAMs modification effectively prevent the aging of perovskite film within three days under dark condition.

4. Conclusion

In this work, we have investigated the performance of perovskite solar cells by inserting a self-assembled monolayer of synthesized triethoxy (4-methoxyphenyl) silane (SAM1) and triethoxy (3,4-dimethoxyphenyl) silane (SAM2) between mesoporous TiO_2 and a perovskite active layer. The synthesized SAMs are characterized using NMR measurements. For a preliminary result, the modified TiO_2 surface has good wetting properties and is more stable in dark conditions. The next plan of this work, we will evaluate the functionality and aging stability of the modified cells.

Acknowledgement
Words cannot express my gratitude to my professor for his invaluable patience and feedback all through the entirety of my research.

Reference

[1] Guang Y, Changlei W, Hongwei L, Xiaolu Z, Pingli Q, Liangbin X, Xingzhong Z, Yanfa Y and Guojia F 2017 *J. Mater. Chem. A* **5** 1658.
[2] Jinyoung H, Hannah K, Eunah K, Dong-Wook K, Hae Jung S and Dong Ha K 2020 *J. Mater. Chem. A* **8** 2105.
[3] Manoso A S, Ahn C, Soheili A, Handy C J, Correia R, Seganish W and DeShong P 2004 *J. Org. Chem* **69** 8305.
[4] Peng J, Duong T, Zhou X, Shen H, Wu Y, Mulmudi H K, Wan Y, Zhong D, Li J, Suzuki T, Weber K J, Catchpole K R and White T P 2017 *Adv. Energy Mater* **4** 1601768.
[5] Shi X, Ding Y, Zhou S, Zhang B, Cai M, Yao J, Hu L, Wu J, Dai S and Nazeeruddin M K 2019 *Adv. Sci (Weinh)* **6** 1901213.
[6] Pilik B and Pratima 2022 *Mater. Res. Bull.* **149** 111700.

IOP Conf. Series: Materials Science and Engineering 1280 (2023) 012013 doi:10.1088/1757-899X/1280/1/012013

Biosynthesis of silver nanoparticles using extract of banana peel waste assisted by microwave irradiation

T Phoemthaisong[1], B Ksapabutr[1], N Chaiyut[1] and M Panapoy[1,2,*]

[1]Department of Materials Science and Engineering, Faculty of Engineering and Industrial Technology, Silpakorn University, Sanamchandra Palace Campus, Nakhon Pathom 73000, Thailand
[2]Center of Excellence on Petrochemical and Materials Technology, Chulalongkorn University, Bangkok 10330, Thailand

*Corresponding author: mpanapoy@gmail.com, panapoy_m@su.ac.th

Abstract. Silver (Ag) nanoparticles are among the most desirable nanomaterials for a variety of applications, such as lithium-ion battery, water pollution monitoring, photocatalytic water splitting, and medical devices. Their increased production and integration in environmental applications play an important role for nanoscience and nanotechnology in engineering. The synthesis methods of Ag nanoparticles include physical, chemical, and biological routes. In this study, we reported a rapid, eco-friendly, and cost-efficient method to synthesize Ag nanoparticles using an extract of banana peel waste as a reducing and capping agent under microwave radiation. A comparison between the conventional heating method and microwave radiation revealed that the microwave radiation method for the extraction of banana peel waste and the synthesis of Ag nanoparticles required much shorter extraction and reaction times. The effects of extraction conditions (banana peel content and extraction time) and synthesis conditions (silver nitrate concentration and reaction time) on Ag nanoparticles were investigated. The optimum conditions were found to be for the extraction: 50 g/L banana peel-to-water ratio and microwave extraction time of 140 s at 800 W and for the Ag nanoparticles synthesis: microwave reaction time of 120 s at 800 W and 1 mM silver nitrate solution and. Plasmonic Ag nanoparticles with a surface plasmon resonance peak at 420 nm and an average size of about 149 nm were synthesized.

1. Introduction

Silver nanoparticles (AgNPs) are significant materials that have received much attention. Because these nanoparticles have unique optical, thermal, electrical, and biological properties, they are used in catalysis, medical engineering, biological imaging, and drug delivery, leading to an increase in their demand [1]. The physical and chemical synthesis of nanoparticles can have significant ecological and economic consequences [2]. Because of their simplicity, eco-friendliness, and cost-effectiveness, biological techniques are regarded as a viable alternative to physical and chemical approaches. Numerous research teams have reported synthesizing AgNPs utilizing plant extracts as both reducing and capping agents [3-6]. Banana peels contain numerous polymers, such as cellulose, hemicellulose, lignin, and pectin, in banana peels that could be utilized to produce silver nanoparticles [7]. Bananas are consumed globally, and the peels are typically thrown away once the pulp has been devoured. Using plant extracts as both reducing and stabilizing agents, microwave-assisted biosynthesis is a feasible approach for the fast and simple production of AgNPs. Microwave irradiation rapidly heats the reaction medium, forming monodispersed nanoparticles in a short reaction time [8]. This work is to study the biosynthesis of AgNPs using banana peel extract as a reducing and capping agent for the synthesis of AgNPs by microwave irradiation.

2. Experimental

2.1 Preparation of banana peel extract (BPE)

The fresh banana peel was washed several times with distilled water to get rid of dust and other organic impurities. In the microwave-assisted method, banana peels were extracted in 100 mL of deionized (DI) water at different ratios of banana peel to water (g/L) in a microwave oven at a constant power of 800 W for various extraction times and filtered through cheesecloth. This filtrate was centrifuged for 10 min at 1000 rpm and twice filtered with Whatman No. 1 filter paper. The resulting filtrate was kept at 4 °C and utilized as a reducing and stabilizing agent. In the microwave synthesis of AgNPs, 10 mL of BPE were added to 40 mL of silver nitrate ($AgNO_3$) solutions with different concentrations. The reaction mixture was then microwave-irradiated in a microwave oven at a constant 800 W power for various reaction times (60, 80, 100 and 120 s). The AgNPs obtained from the microwave extraction and synthesis were designated as (E)meX(S)mrT, where (E) is the ratio of peel to water for extraction (10, 50, 100, and 150 g/L), meX is the microwave extraction time (60, 100, 140, and 180 s), (S) is the concentration of $AgNO_3$ solution (0.5, 1.0, 1.5, and 2.0 mM), and mrT is the microwave reaction time (60, 80, 100, and 120 s).

In the conventional heating method, 5 g of banana peel was placed in a 250-mL conical flask with 100 mL DI water before being heated for 30 min at 80 °C for extraction and filtered through cheesecloth. After extraction, 10 mL of BPE were added to 40 mL of 1 mM $AgNO_3$ solution. After that, the mixture was boiled at 60 °C for 60 min in AgNPs synthesis. The sample was notated as (E)ceX(S)crT or (50g/L)ce30min(1.0mM)cr1h, where (E) is the ratio of peel to water (50 g/L), ceX is the conventional heating extraction time (30 min), (S) is the $AgNO_3$ solution concentration (1 mM), and crT is the conventional heating reaction time (1 h).

2.2 Characterization of the AgNPs

The characterization of AgNPs was carried out using the UV-3600i Plus Ultraviolet-Visible-Near Infrared Spectrophotometer (UV-VIS-NIR) from Shimadzu. The reduction of Ag^+ ions with BPE in solution was monitored in the wavelength range of 200-800 nm. The AgNPs particle size was evaluated using dynamic light scattering with a Zetasizer Nano-ZS from Malvern.

3. Result and discussion

The UV–vis absorption spectra of AgNPs are shown in Figure 1. The absorption characteristic maximum of AgNPs revealed a surface plasmon resonance (SPR) band at approximately 420 nm. Figure 1(a) shows the UV–vis absorption spectra of a comparison between the conventional heating method and microwave radiation, which revealed that the microwave radiation method for the extraction of banana peel waste and the synthesis of AgNPs required much shorter extraction and reaction times. The microwave radiation technique involves fast initial heating and the creation of localized heat at the reaction sites, resulting in an increment in the reaction rate and, consequently, the productivity of tiny nanoparticles [9]. The results mentioned above show that AgNPs synthesis by microwave radiation has better efficiency than that by conventional heating. As a result, the effect of microwave heating on AgNP synthesis was investigated. The UV–vis absorption spectra of AgNPs became different as $AgNO_3$ concentration increased (0.5 mM to 2 mM), and the maximum peak intensity was reached at 2 mM $AgNO_3$, as shown in Figure 1(b). The broadening of the peak breadth and bump for increasing concentrations suggest the aggregation of smaller nanoparticles, which increased the nanoparticle size [10]. The peak intensity of 1 mM $AgNO_3$ exhibited a sharp, narrow peak of absorbance. As the sharpness of the absorbance peak is dependent on the produced nanoparticle size, this was an indication of the creation of substantially smaller AgNPs [11]. After optimizing the reaction conditions, the optimal concentration of $AgNO_3$ was determined to be 1 mM; consequently, this concentration was selected for all subsequent studies of the reaction parameters. Figure 2(a) shows the absorbance of the SPR band of the AgNPs solutions irradiated for various reaction times (at 60, 80, 100, and 120 s). The maximum absorbance increased in intensity as a function of irradiation time. These results suggested that the content of AgNPs increased as irradiation time increased [12]. At 120 s of reaction time, the maximum peak intensity was obtained. Figure 2(b) shows the absorbance of SPR band of other AgNPs solutions obtained with different BPE concentrations (10, 50, 100, and 150 g/L). The intensity of absorbance peak increased rapidly as the BPE concentration rose from 10 to 50 g/L. Increased hydroxyl groups within the BPE promoted the Ag^+ complexation with the molecular matrix [13]. However, the peak intensity decreased as the BPE content was raised from 100 to 150 g/L, indicating that the number of particles had decreased. The biomolecules in BPE and $AgNO_3$ may have reached a saturation limit, which suggests that the concentrated biomolecules impeded the reduction of the $AgNO_3$ precursor and lowered the particle production rate [13]. Figure 2(c) shows the absorbance of SPR band of the AgNPs solutions irradiated for various extraction times (at 60, 100, 140, and 180 s). The greatest peak intensity was achieved at 140 s of extraction, and the maximum absorbance increased in intensity as irradiation time increased. The above results showed that for the extraction by microwave radiation approach, the optimal conditions were a 50 g/L banana peel-to-water ratio and a microwave extraction time of 140 s at 800 W, whereas for the AgNPs synthesis, the optimal parameters were a 1 mM $AgNO_3$ solution and a microwave reaction time of 120 s at 800 W.

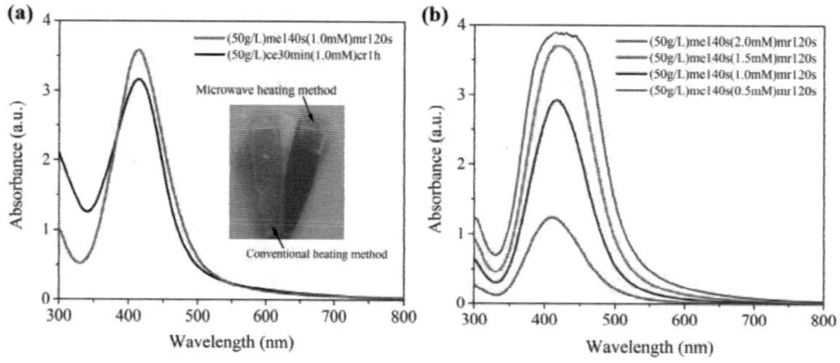

Figure 1. The UV–vis absorption spectra of the prepared AgNPs: (a) a comparison between the conventional and microwave techniques (the insert is the corresponding sample) and (b) various concentrations of $AgNO_3$ for the microwave method.

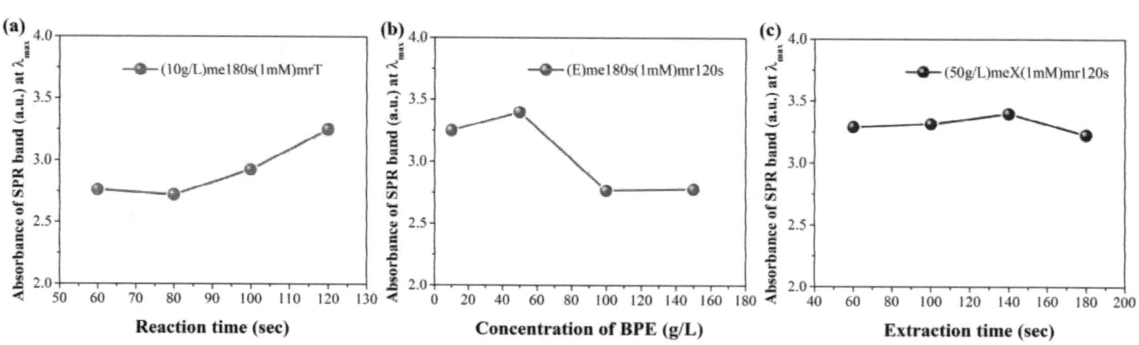

Figure 2. The absorbance of SPR band of the synthesized AgNPs at different (a) reaction times, (b) concentrations of BPE, and (c) extraction times.

The average size of AgNPs in solution was studied by the DLS technique. Figure 3(a) shows the average size of the AgNPs produced by the microwave radiation method; the position of peak 1 is 197 nm, the position of peak 2 is 4,119 nm, and the average size is 149 nm and a polydispersity index (PDI) is 0.348, which is considered as moderate polydispersity [14]. Figure 3(b) depicts the average size of AgNPs synthesized by conventional heating method; peak 1 is located at 331 nm, peak 2 is located at 4714 nm, and the average size is 238 nm with a PDI of 0.358, indicating moderate polydispersity [14]. Both methods exhibited that a greater proportion of the particles were found in smaller agglomerates of AgNPs and a smaller proportion in larger agglomerates of AgNPs. These results are consistent with other studies for the formation of AgNP agglomeration [15]. The above results show that AgNPs synthesized by microwave radiation are smaller than AgNPs synthesized by conventional heating, which is consistent with the UV-vis analysis in Figure 1(a).

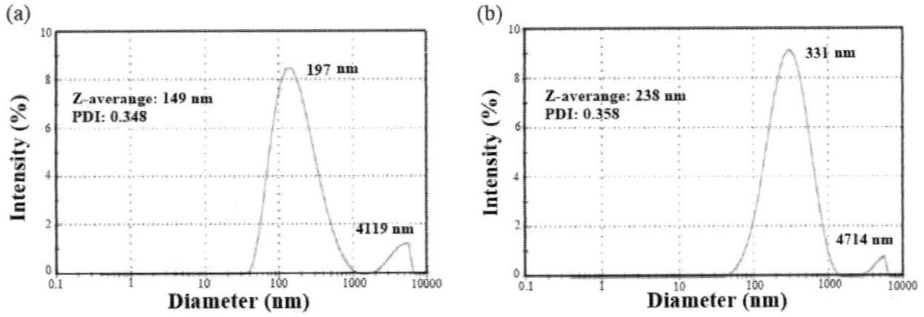

Figure 3. DLS spectra of AgNPs produced by: (a) microwave heating method and (b) the conventional heating method.

4. Conclusions

Banana peel waste was successfully utilized to synthesize AgNPs with the microwave irradiation method. A comparison between the microwave irradiation and conventional heating method revealed that the microwave radiation method for the extraction of banana peel waste and the synthesis of AgNPs required much shorter extraction and reaction times. The optimum conditions were found to be for the extraction: 50 g/L banana peel-to-water ratio and microwave extraction time of 140 s at 800 W; and for the synthesis of AgNPs: 1 mM $AgNO_3$ solution and microwave reaction time of 120 s at 800 W. The average diameter of AgNPs synthesized by the microwave radiation method is smaller than that of AgNPs prepared by the conventional heating technique.

Acknowledgement

The authors thank the Department of Materials Science and Engineering, Faculty of Engineering and Industrial Technology, Silpakorn University.

References

[1] Jain P K, Huang X, El-Sayed I H and El-Sayed M A 2008 *Acc. Chem. Res* **41** 1578.
[2] Gopinath V, MubarakAli D, Priyadarshini S, Priyadharsshini N M, Thajuddin N and Velusamy P 2012 *Colloids surf. B* **96** 69.
[3] Shankar S S, Rai A, Ahmad A and Sastry M 2004 *J. Colloid Interface Sci.* **275** 496.
[4] Bar H, Bhui DK, Sahoo G P, Sarkar P, De S.P and Misra A 2009 *Colloids Surf.* **339** 134.
[5] Thirunavoukkarasu M, Balaji U, Behera S, Panda P K and Mishra B K 2013 *J. Spectrochim. Acta A.* **116** 424.
[6] Vidhu V K and Philip D 2015 *Environ. Nat. Resour. J* **6** 1.
[7] Emaga T H, Andrianaivo R H, Wathelet B, Tchango J T and Paquot M 2007 *Food Chem.* **103** 590.
[8] Nadagouda M N, Speth T F and Varma R S 2011 *Acc. Chem. Res* **44** 469.
[9] Tu W and Liu H J 2000 *J. Mater. Chem* **10** 2207.
[10] Parit S B, Karade V C, Patil R B, Pawar N V, Dhavale R P, Tawre M, Pardesi K, Jadhav U U, Dawkar V V, Tanpure R S, Kim J H, Jadhav J P and Chougale A D 2020 *Mater. Today Chem.* **17** 100285.

[11] Philip D 2010 *Physica E Low Dimens. Syst. Nanostruct* **42** 1417.
[12] Carbone K, Angelis A D, Mazzuca C, Santangelo E, Macchioni V, Cacciotti I, Petrella G.Cicero D O and Micheli L 2020 *LWT* **133** 110088.
[13] Peng H, Yang A and Xiong J 2013 *Carbohydr. Polym* **91** 348.
[14] Bhattacharjee S 2016 *J. Control. Release* **235** 337.
[15] Hudecová A, Kusznierewicz B, Pran E, Magdolenová Z, Hašplová K, Rinna A, Fjellsbø L, Kruszewski M, Lankoff A, Sandberg W, Refsnes M, Skuland T, Schwarze P, Brunborg G, Bjøras M, Collins A, Miadoková E, Gálová E, Dušinská E and Dušinská M 2012 *Mutagenesis* **27** 759.

IOP Conf. Series: Materials Science and Engineering 1280 (2023) 012014 doi:10.1088/1757-899X/1280/1/012014

Dry Electropolishing of Laser Powder Bed Fusion Manufactured Cobalt-chrome

T Rautio*, M Jaskari, J Päkkilä and A Järvenpää

Future Manufacturing Technologies, University of Oulu, Pajatie 5, 85500 Nivala, Finland.

*Corresponding author: timo.rautio@oulu.fi

Abstract. Cobalt-chrome (Co-Cr) has been used extensively in medical applications, gas turbines and dental products for decades now. Laser powder bed fusion (PBF-LB) manufacturing is very suitable for small series and prototyping and in this regard its application to dental products is promising as they are usually unique and high precision is required. One major factor limiting the application of the PBF-LB technique is the surface quality, which is not adequate in as built condition. This work focuses on improving the surface quality of the PBF-LB manufactured Co-Cr with dry electropolishing so that it is suitable for dental bridges. The surface quality is analysed using laser scanning microscopy for surface roughness, surface profile and topography to reveal the surface characteristics after the printing and the effect of dry electropolishing is highlighted. The results showed that dry electropolishing can be used successfully to improve the surface quality for the needs of dental bridges. Final product quality was verified by a dentist, and the dental bridge could be used on a patient.

1. Introduction

Cobalt-chromium (Co-Cr) alloys have been used for partial denture prosthesis for several decades now [1]. In the early days, those were manufactured by casting, but new emerging technologies such as the laser powder bed fusion (PBF-LB) for additive manufacturing (AM) can provide several advantages over those. This technique is very suitable for small products with complex structures that have high customization such as the dental products, which are identical for each patient [2]. Hong et al. showed that better mechanical properties including tensile strength, yield strength and hardness can be obtained with the AM technique compared to casting [3].

Some studies can be found in the open literature about the usage of PBF-LB manufactured Co-Cr in dental applications. Zhang et al. found that there was no significant difference between cast and PBF-LB manufactured clasps in terms of fitness accuracy [4]. McDonald used laser polishing successfully to improve the surface quality of Co-Cr dental implants manufactured with PBF-LB [5]. In this work, we focus on the dry electropolishing of PBF-LB Co-Cr in combination with glass bead blasting to gain improved surface quality for dental applications.

2. Material and Methods

2.1 Specimen manufacturing and the material

A Concept Laser printer (Mlab cusing) based on the PBF-LB manufacturing technique was used for all printing in this study. The print chamber size of the machine was 90 mm x 90 mm x 80 mm. The used Co-Cr powder was Remanium® Star CL, which was also provided by Concept Laser and had the chemical composition detailed in Table 1. This powder was specifically designed for dental applications and gives a yield strength of 822 MPa, ultimate tensile strength of 1156 MPa and an elongation of 8 %. Most relevant printing parameters were as follows: power (*P*) of 95 W, laser speed (*v*) of 650 mm/s, hatch spacing (*h*) of 0.11 mm and layer thickness (*t*) of 25 µm. Heat treatment was carried out in argon atmosphere and it was carried out as follows: preheating to 600 °C, heating to 1150 °C with maximum rate, hold for 1h and finished with a cool down to 600 °C with the maximum rate. Cooling down to room temperature was then allowed in air.

Table 1. Chemical composition of the cobalt-chrome powder in wt-%.

Co	Cr	W	Si	Mn	N	Nb	Fe
60.5	28	9	1.5	<1	<1	<1	<1

2.2 Dry electropolishing and characterization of the surfaces.

Dry electropolishing was conducted with a DLyte HF100+ machine with polymer medias (CoCr-01 and CoCr-06) supplied by the manufacturer. The used machine parameters are listed in Table 2 showing the three-step polishing cycle. First two steps were used for roughness reduction using CoCr-01 media. The final third step was used for surface brightening with CoCr-06 media, which consists of smaller particles. Surface roughness and the topographies of the Co-Cr samples in as built and dry electropolished conditions were measured with a Keyence VK-X200 series laser microscope. Glass bead blasting was performed before the dry electropolishing.

Table 2. Used DLyte parameters for the dry electropolishing experiments.

Media	Voltage [V]	T1 [µs]	T2 [µs]	T3 [µs]	T4 [µs]
CoCr-01	28	500 000	3 000 000	500 000	4 000 000
CoCr-01	30	50 000	300 000	50 000	500 000
CoCr-06	30	10	10	30	10

3. Results and discussion

3.1 Surface roughness and profile of the PBF-LB manufactured Co-Cr

The surface roughness and profiles of the PBF-LB manufactured Co-Cr in as built and dry electropolished conditions were measured utilizing confocal laser optical microscopy. The resulting surface roughness values are presented in Table 3. The R_a value was decreased by 84% from 3.89 to 0.61 µm and the R_z had also a notable 73% decrease. The surface profile of the as built condition is presented in Fig. 1a) and it shows the rough surface after the printing. The surface has a wavy appearance with some higher frequency roughness on top. A few sharper notches are also visible. The results after the dry electropolishing show a clear improvement (Fig. 1b). Polishing was able to remove all the high frequency components in the profile and only the subtle wavy appearance is left. Peak-to-peak variation is also clearly smaller as also evidenced by the Rz measurements.

Table 3. Surface roughness of the laser powder bed fusion manufactured cobalt-chrome and the effect of dry electropolishing.

State	Ra (μm)	Rz(μm)
As built	3.89	34.2
Dry electropolished	0.61	9.11

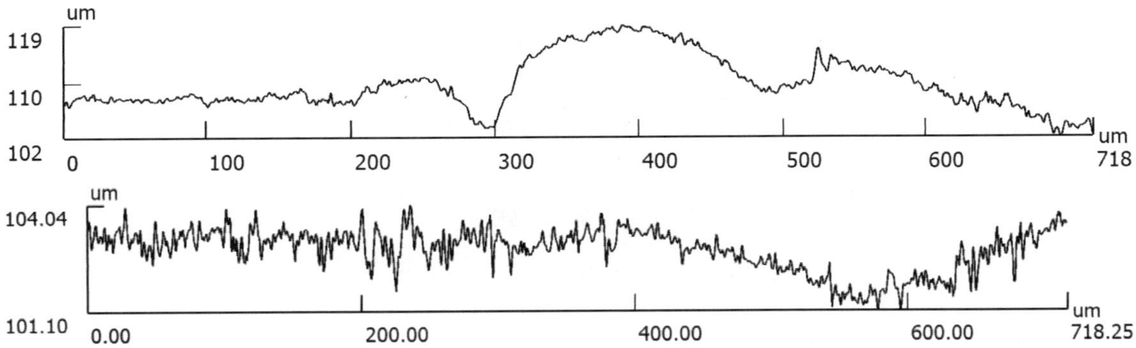

Figure 1. Surface profile of the laser powder bed fusion manufactured Co-Cr in a) as built and b) dry electropolished condition.

3.2 Surface topography of the Co-Cr before and after dry electropolishing

Laser optical microscopy was also utilized for the measurement of the surface topography of both conditions. The as built surface presented in Fig. 2a) shows the rough appearance of the surface when no post processing has been applied. The surface has a mix of higher peaks with the diameter ranging from a few tens of microns to roughly 200 microns. Some of the formations are related to partially unmelted powder particles and their conglomerates.

The topography after the glass bead blasting and dry electropolishing presented in Fig. 2b) shows clearly smoother appearance compared to the as built surface. As shown with the roughness measurements and profiles, the peak-to-peak deviations are halved from the as built surface. There are only a couple of higher peaks present in the surface and they have multiple times larger diameter compared to the as built surface. The high scale of the remaining waves appearing on the surface and the removal of abrupt changes have resulted in a mirror like finish of the surface

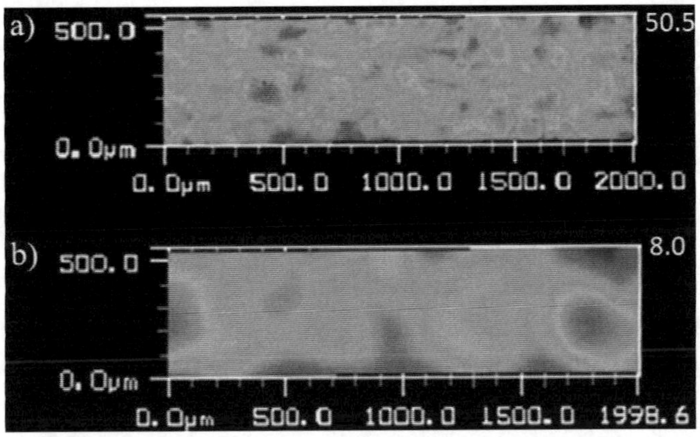

Figure 2. Surface topography of the laser powder bed fusion manufactured Co-Cr in as built and b) dry electropolished condition.

3.3 Application of the dry electropolishing on dentures

The dry electropolishing method was also used on actual parts, namely partial dentures. These specimens were manufactured with the same parameters as the samples for surface characterization. The resulting denture after the dry electropolishing is presented in Fig. 3. Another piece was inspected by a dental professional and it was accepted for use with a patient.

Figure 3. Photo of a laser powder bed fusion manufactured cobalt-chrome partial denture after dry electropolishing.

4. Conclusions

This work investigated the dry electropolishing of laser powder bed fusion (LPBF) manufactured cobalt-chrome and can be concluded as follows:

- The surface roughness could be decreased from 34.2 to 9.11 μm in terms of R_z and 84% decrease in R_a to a value of 0.61 μm was measured.
- Polishing was able to remove the rough features of the surface such as the partially melted particles to give a smooth and mirror polished appearance.
- Sufficient surface quality was achieved for a polished partial denture so that it could be used with a patient.

Acknowledgement

The authors would like to acknowledge the financial support received from the Council of Oulu Region and the European Union (European Regional Development Fund) for the M3D project and Business Finland for the DREAMS project.

References

[1] Peyton F A 1958 *Dent. Clin. N. Am.* **2** 759.
[2] Savolainen J and Collan M 2020 *Addit. Manuf.* **32** 101070.
[3] Hong J H and Yeoh F Y 2020 *Mater. Today: Proc.* **29** 196.
[4] Zhang M, Gan N, Qian H and Jiao T 2022 *J. Prosthodont. Res.* **66** 459.
[5] McDonald M W, Gora W S, Stevenson S G, Weston N J and Hand D P 2020 *J. Laser Appl.* 042019.

Preparation and Properties of Carboxymethyl Cellulose / β-Cyclodextrin Hydrogel

N Panyakaew[1], C Thongpin[1] and S Hemsri[1]*

[1]Department of Materials Science and Engineering, Faculty of Engineering and Industrial Technology, Silpakorn University, Sanamchandra Palace Campus, Nakhon Pathom 73000, Thailand

*Corresponding author: sudsirihemsri@yahoo.com

Abstract. Carboxymethyl cellulose (CMC) based hydrogels provide superabsorbent properties but show low mechanical properties. In order to overcome this problem, β-cyclodextrin (CD) is added to improve the mechanical strength of hydrogels. Moreover, CD-containing hydrogels have the potential to encapsulate antimicrobial essential oils or drugs and give a sustained release for medical applications to treat a variety of bacterial infectious diseases causing destruction to human health. In this work, the CMC/CD hydrogels were prepared by using epichlorohydrin as a chemical crosslinking agent. The objective of this research was to study the effect of the CMC/CD weight ratio at 6:1, 7:1, and 8:1 on morphology, mechanical properties and swelling ratio in water of the CMC/CD hydrogels compared with CMC hydrogel. The weight ratio of epichlorohydrin and CMC/CD was constant at 6:1. The morphology analysis of the CMC/CD hydrogels showed an irregular porous structure with the presence of -like structure. The addition of CD content led to an increase in compressive strength whereas the swelling ratio of the hydrogel decreased. The CMC/CD hydrogel at weight ratio of 7:1 provided the highest compressive strength and also retained swelling ability.

1. Introduction

Hydrogels are hydrophilic polymers with three-dimensional crosslinked polymeric networks. They are mainly synthesized from polymers with hydrophilic groups such as carboxyl, hydroxyl, amino and sulfonic acid groups in the polymer chains [1]. These hydrophilic groups in the polymer chain of the hydrogels can absorb water and hold large amounts of water within their porous structure of three-dimensional crosslinked networks [2]. The water sorption ability and swelling kinetics of a hydrogel depend on its porosity [3]. Hydrogels are insoluble in water at physiological temperature and pH because of the presence of chemical or physical crosslinks but swell considerably in an aqueous medium. Due to their high-water absorption capacity and biocompatibility, hydrogels are widely used in extensive applications, especially biomedical applications such as tissue engineering, contact lenses, wound dressings, and drug delivery [2].

Carboxymethyl cellulose (CMC) is a cellulose derivative with many hydroxy and carboxyl groups and can absorb water. Although the CMC-based hydrogels have excellent biodegradation and superabsorbent properties, they have low mechanical properties. Therefore, the mechanical properties of CMC-based hydrogels can be improved by multi-component hydrogel systems [3-5].

Cyclodextrins have been used to encapsulate hydrophobic compounds. The β-Cyclodextrin (CD) is a torus-shaped cyclic oligosaccharide consisting of 7 monomers of D-glucose linked through α-1,4 glycosidic bonds and they have unique structure with hydrophobic internal surface and hydrophilic outer surface. CD-containing hydrogels have potential to encapsulate antimicrobial essential oils or drugs for medical applications and give a sustained release to treat a variety of bacterial infectious diseases causing destruction to human health [6-8].

In this study, the CMC/CD hydrogels were prepared by using epichlorohydrin as a chemical crosslinking agent. The objective of this research was to study the effect of the CMC/CD weight ratio at 6:1, 7:1, and 8:1 on morphology, mechanical properties and swelling ratio in water of the CMC/CD hydrogels compared with CMC hydrogel.

2. Experimental

2.1 Materials

Carboxymethyl cellulose (CMC) (Mw = 90,000 g/mol with substitution 0.7, β-Cyclodextrin (CD) (purity ≥97%) and epichlorohydrin (purity ≥97%) were purchased from Sigma–Aldrich. Sodium Hydroxide (NaOH) (assay 97%) was obtained by Ajax FineChem.

2.2 Preparation of hydrogels

The CMC and CD at various weight ratios of 1:0, 6:1, 7:1 and 8:1 were completely dissolved in 6%wt NaOH solution at 60 °C with stirring speed of 600 rpm. Epichlorohydrin as a crosslinking agent was added dropwise to the stirred CMC/CD solution at 60 °C with stirring speed of 600 rpm. The crosslinking reaction was carried out at 60 °C for 5 h and highly viscous gel was formed. The gel was soaked in reverse osmosis water several times until the gel reached neutral pH. Finally, the gel was dried in vacuum oven at 60 °C for 48 h.

2.3 Characterization

2.3.1 Field Emission-Scanning Electron Microscope (FE-SEM)

The morphology of hydrogels was observed using FE-SEM (Tescan model Mira3 from Czech Republic). Before SEM investigation, the dried hydrogels were swollen in deionized water at room temperature for 24 h and then freeze-dried for 21 h. The freeze-dried samples were coated with a thin layer of gold before SEM observation. The surface morphologies of hydrogels were observed at an acceleration voltage of 10 kV.

2.3.2 *Swelling characteristic*

The dried hydrogels were immersed in deionized water at room temperature for 40 h. The swelling test was carried out in triplicate. Water-swelling ratios were calculated using the following equation:

$$\text{Swelling ratios} = (W_t - W_d)/W_d$$

where W_t is the weight of the swollen hydrogel and W_d is the initial weight of the dried hydrogel. The hydrogel was removed from deionized water before weighing, and the excess of water on the hydrogel surface was gently absorbed with tissue towel paper.

2.3.3 Compressive test

Compressive tests were performed by Universal Testing Machine (Instron 3366, Instron Engineering corporation). The hydrogel samples were cut with a 40 mm diameter and a 10 mm height. The samples were compressed until the compression ratio reached 80% at a rate of 20 mm/min.

3. Results and discussion

3.1 Structural Morphologies of the CMC and CMC/CD hydrogels

The microstructure of the swollen CMC and CMC/CD hydrogels after freeze-drying was observed by using the FE-SEM. The cross-sectional morphologies of all hydrogels at 100X and 500X magnification are illustrated in Figure 1.

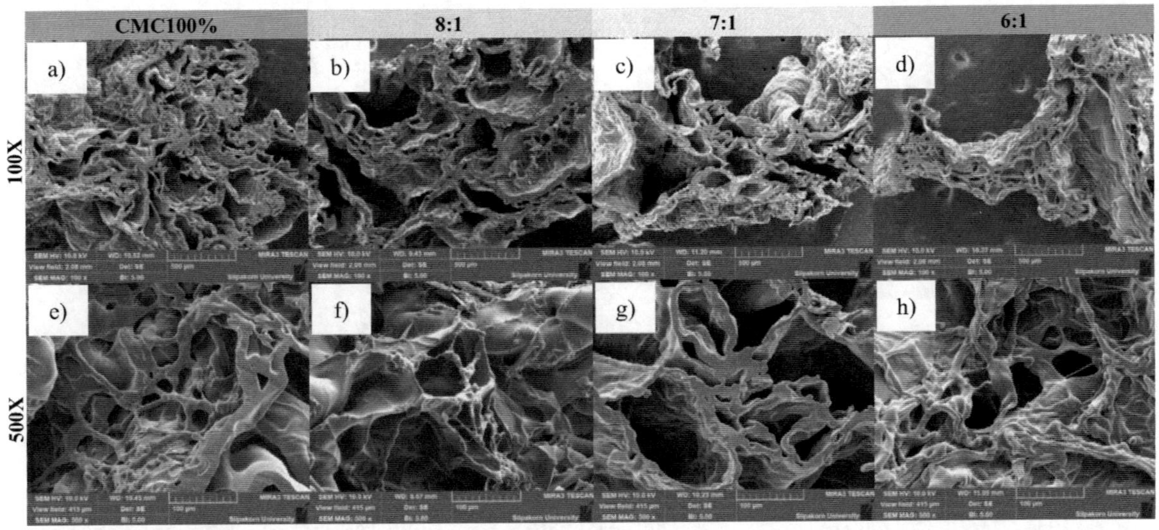

Figure 1. FE-SEM images of CMC hydrogel (a,e), CMC/CD hydrogel at ratio of 8:1 (b,f), CMC/CD hydrogel at ratio of 7:1 (c,g) and CMC/CD hydrogel at ratio of 6:1 (d,h) at 100X and 500X magnification.

The microstructures of CMC and CMC/CD hydrogels showed irregular pours structures with the presence of open channel and partial fiber-like structures within the hydrogel. At higher magnification, among CMC/CD hydrogels, the CMC/CD hydrogel at weight ratio of 7:1 (as represented in Fig.1g) had thicker pore walls compared with the others, resulting in higher compressive strength as it will be discussed later in the next section of 3.3.

3.2 Swelling behavior of the CMC and CMC/CD hydrogels

Swelling is one of properties indicating water holding ability and permeability of hydrogels. During swelling process, water diffuses into the hydrogel network and forms hydrogen bonds with negatively charged groups and polar hydrophilic moieties such as hydroxy groups in the polymer chain of the hydrogels, until free water molecules which absorbed within the network attain equilibrium [9].

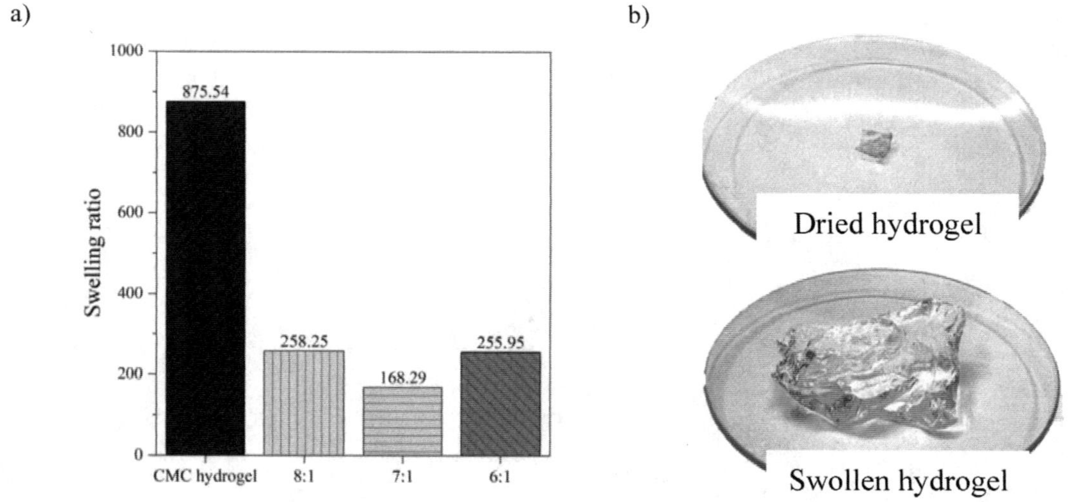

Figure 2. Swelling ratio of CMC and CMC/CD hydrogels (a) and dried and swollen hydrogels (b).

Figure 2 showed the result of swelling ratio of the CMC and CMC/CD hydrogels. According to Fig.2, the CMC hydrogel had the highest swelling ratio due to the presence of electrostatic repulsion between carboxylate anions (COO^-) in the hydrogel which could produce more hydrophilicity and drive hydrogel networks to expand, resulting in high water swelling ability in the CMC hydrogel [10]. On the other hand, the CMC/CD hydrogels provided a lower swelling ratio with respect of CMC hydrogel because the CD has hydrophobicity and rigid structure, leading to a decrease in flexibility and hydrogel network expansion [3].

3.3 Compressive property of the CMC and CMC/CD hydrogels

The compressive test was used to investigate the mechanical characteristics of the hydrogels. Table 1 displays the compressive strength of the CMC, and CMC/CD hydrogels at weight ratios of 8:1, 7:1, and 6:1 with values of 5.99, 19.86, 24.17, and 17.25 kPa, respectively. The results showed that the CMC/CD hydrogels provided higher compressive strengths compared to the CMC hydrogel. This might be due to the steric effect of the CD molecule incorporated in polymer networks of hydrogels and its rigidity of cyclic CD structure. Thus, flexibility of the CMC/CD hydrogel network decreased and the CMC/CD hydrogels were harder than the CMC hydrogel, leading to higher compressive strength [5]. Among the CMC hydrogels with containing CD, the highest compressive strength was found in the CMC/CD at 7:1 weight ratio. This result could be explained by the presence of a thicker pore wall in the CMC/CD hydrogel structure (as shown in the SEM image in Fig.1g), leading to a strong structure in the hydrogel, which supported against compressive force. A similar result was reported by Sringam et.al [11]. In this work, it was found that starch hydrogel with a thicker pore wall thickness could enhanced compressive strength.

Table 1. Compressive strength of CMC hydrogel and CMC/CD hydrogels.

Formula	**CMC hydrogel**	**CMC/CD 8:1**	**CMC/CD 7:1**	**CMC/CD 6:1**
Compressive strength (kPa)	5.99±1.77	19.86±1.14	24.17±1.53	17.25±2.41

4. Conclusions

In this research, the morphology analysis of the CMC and CMC/CD hydrogels showed an irregular porous structure with the presence of fiber-like structure. The CMC/CD hydrogel showed highly enhanced mechanical strength while the swelling properties significantly decreased. The CMC/CD hydrogel at weight ratio of 7:1 provided the highest compressive strength and also retained swelling ability. The fabricated CMC/CD hydrogels in this work have possible potential to encapsulate hydrophobic essential oils or drugs for medical applications.

Acknowledgement

The authors gratefully acknowledge financial support from Department of Materials Science and Engineering, Faculty of Engineering and Industrial Technology, Silpakorn University and Graduate School Silpakorn University.

References

[1] Chamkouri H and Chamkouri M 2021 *Am. J. Biomed. Sci. & Res.* **11** 485.
[2] Bharskar GR 2020 *World J. Pharm. Pharm. Sci.* **9** 1288.
[3] Mondal MdIH 2018 (Eds.) In Cellulose-Based Superabsorbent Hydrogels, Polymers and Polymeric Composites: A Reference Series, Springer International Publishing, 1 pages.
[4] Adel AM, Abou-Youssef H, El-Gendy AA and Nada AM 2010 *Nature and Science* **8** 244.
[5] Jeong D, Jooa S-W, Hua Y, Shindeb VV, Chob E and Junga S 2018 *Eur. Polym. J.* **105** 17.
[6] Liu J, Tian B, Liu Y, Wan JB 2021 *Int. J. Mol. Sci.* **22** 13516.
[7] Piletti R, Bugiereck AM, Pereira AT, Gussati E, Dal Magro J, Mello JMM, Dalcanton F, Ternus RZ, Soares C, Riella HG and Fiori MA 2017 *Mater. Sci. Eng. C Mater. Biol. Appl.* **75** 259.
[8] Kono H, Onishi K and Nakamura T 2013 *Carbohydr. Polym.* **98** 784.
[9] Malik NS, Ahmad M and Minhas MU 2017 *PLoS One* **12** 1.
[10] Liu YY and Fan XD 2005 *Biomaterials* **26** 6367.
[11] Sringam J, Pankongadisak P, Trongsatitkul T and Suppakarn N 2022 *Polymers* **14** 3552.

IOP Conf. Series: Materials Science and Engineering 1280 (2023) 012016 doi:10.1088/1757-899X/1280/1/012016

Microwave-assisted green synthesis of silver nanoparticles using extract of mint leaves

A Thanasakonpong[1,2], B Ksapabutr[1] and M Panapoy[1,2,*]

[1]Department of Materials Science and Engineering, Faculty of Engineering and Industrial Technology, Silpakorn University, Sanamchandra Palace Campus, Nakhon Pathom 73000, Thailand
[2]Center of Excellence on Petrochemical and Materials Technology, Chulalongkorn University, Bangkok 10330, Thailand

*Corresponding author: mpanapoy@gmail.com, panapoy_m@su.ac.th

Abstract. Silver nanoparticles are widely used in areas such as biosensing, hydrogen production, electronics, photovoltaics, antimicrobials, and biomedical engineering applications. Biological synthesis is regarded as a safe and non-toxic process, but its main disadvantage is the slow processing time. The mint leaf extract can act as both a reducing and stabilizing agent for the biological synthesis of silver nanoparticles. In this work, a rapid and green method for the microwave-assisted biological synthesis of silver nanoparticles was investigated using an aqueous leaf extract of mint as a biosource of cost-effective, non-hazardous reducing and stabilizing agents. The optimum conditions for microwave extraction of mint leaves were 220 s at 800 W and a 5 g/L mint-to-water ratio. Meanwhile, the optimal conditions for the biological synthesis of silver nanoparticles were as follows: concentration of silver nitrate solution = 2 mM, microwave reaction time = 100 s at 800 W, and pH = 8. The silver nanoparticles obtained by the microwave heating process showed a surface plasmon band centered at 420 nm with a higher peak height than those prepared by the conventional heating process. Dynamic light scattering analysis has shown the synthesized silver nanoparticles to have an average diameter of 54 nm.

1. Introduction

Nanoparticles are one of the most rapidly growing nanomaterials due to their unique physical, chemical, and biological characteristics. They are often used for different purposes, such as medicinal applications, catalytic reactions, and solar cells. Although there are many different types of nanoparticles, silver nanoparticles (AgNPs) have attracted a lot of attention from researchers and industry because of their optical properties as a result of their plasmonic effect [1-2]. AgNPs synthesis has been the subject of extensive research and development over the years. One of the production processes for AgNPs uses plant extracts as a reducing and stabilizing agent and produces yields comparable to chemical synthesis. Therefore, plant extracts provide a new alternative to traditional reducing agents such as sodium borohydride ($NaBH_4$). Moreover, the colloidal stability of plant extracts has been reported to be superior to citrate-capped AgNPs. Mint leaf is a plant whose main constituents are phenolic compounds such as polyphenols, alkaloids, carbohydrates, etc. The main advantage of using mint leaf extract (MLE) is that it is renewable and non-toxic to the environment [2-4]. In this work, a rapid and green method for the microwave-assisted biological synthesis of AgNPs was investigated using an aqueous leaf extract of mint as a reducing and stabilizing agent.

2. Experimental

2.1 Preparation of mint leaf extract (MLE) and synthesis of AgNPs

For the microwave heating process (MHP), finely chopped fresh mint leaves (0.1, 0.5, 1.0, and 2.0 g) were extracted with 100 mL of deionized water under microwave irradiation at a constant power of 800 W for 60, 100, 140, 180, and 220 s and filtered. After extraction, the filtrates were adjusted to pH 7, 8, and 9. In AgNPs synthesis, 10 mL of aqueous solutions of $AgNO_3$ (0.5, 1.0, 1.5, and 2.0 mM) were added to 40 mL of the MLE and then placed in the microwave at 800 W for 60, 80, 100, 120, and 140 s. The samples were named as (E)meX(S)mrY, where (E) is the mint-to-water ratio, meX is the microwave extraction time, (S) is the concentration of $AgNO_3$ solution, and mrY is the microwave reaction time (Table 1).

For the conventional heating process (CHP), 0.5 g of finely chopped fresh mint leaves were extracted with 100 mL of deionized water at 60 °C for 1 h and filtered. The filtrates were adjusted to pH 8. In a typical synthesis, 10 mL of aqueous solutions of $AgNO_3$ (2 mM) were added to 40 mL of the MLE and maintained at 60 °C for 1 h. This sample was labelled as (E)ceX(S)crY, where (E) is the mint-to-water ratio, ceX is the conventional heating extraction time, (S) is the concentration of $AgNO_3$ solution, and crY is the conventional heating reaction time (Table 1).

Table 1 The sample abbreviations used in this work.

Sample	(E) (g/L)	meX (s)	ceX (h)	pH	(S) (mM)	mrY (s)	crY (h)
(5g/L)me220s	5	220	-	8	-	-	-
(5g/L)ce1h	5	-	1	8	-	-	-
(5g/L)ce1h(2mM)cr1h	5	-	1	8	2.0	-	1
(E)me220s(2mM)mr100s	1, 5, 10, 20	220	-	8	2.0	100	-
(5g/L)me220s(2mM)mr100s	5	220	-	7, 8, 9	2.0	100	-
(5g/L)meX(2mM)mr100s	5	60, 100, 140, 180, 220	-	8	2.0	100	-
(5g/L)me220s(S)mr100s	5	220	-	8	0.5, 1.0, 1.5, 2.0	100	-
(5g/L)me220s(2mM)mrY	5	220	-	8	2.0	60, 80, 100, 120, 140	-

2.2 Nanoparticle characterization

The formation of AgNPs was confirmed using a Ultraviolet-Visible-Near Infrared (UV-VIS-NIR) spectrophotometer (UV-3600i Plus, Shimadzu) in the 200-800 nm wavelength range. The size distribution profile, polydispersity index (PDI), and particle size distribution of the synthesized AgNPs were determined by Dynamic Light Scattering (DLS) analysis using a Zetasizer Nano ZS instrument (Malvern).

3. Results and discussion

A comparison of the heating process for the extraction of mint leaves and the synthesis of AgNPs is shown in Figure 1. Both processes have the same conditions: the mint-to-water ratio, pH, and concentration of $AgNO_3$ solution. Figure 1(a) shows the UV absorbance of MLE obtained by the MHP process compared with the CHP process at pH 8. The UV-visible spectra of both MLEs show the bands at around 260 nm and 320 nm, corresponding to polyphenols [5-6]. It can be seen that the absorbance of MLE prepared by MHP was higher than that of CHP. The results suggested that microwaves contributed to more efficient extraction of polyphenols from mint leaves, probably because microwave radiation causes a sudden increase in pressure and temperature inside plant cells. As a result, the cell wall breaks down and easily diffuses into the solvent [7]. Figure 1(b) shows the surface plasmon resonance (SPR) of AgNPs compared between MHP and CHP processes. The experimental results indicated that the absorbance from the MHP was significantly superior [8]. This is because MLE produced by MHP contains more polyphenols, contributing to the reduction of Ag^+ to Ag^0, than MLE produced by CHP. Various effects on the microwave synthesis of AgNPs have been studied by UV-visible spectroscopy. Figure 2(a) shows the influence of the mint-to-water ratio on the formation of AgNPs. It was found that the MLE at 5 g/L offered higher absorbance than other ratios. At higher mint-to-water ratios, broad SPR bands are observed due to large anisotropic nanoparticles [9-11]. Figure 2(b) shows the influence of pH on AgNPs synthesis. An increase in pH from 7 to 8 resulted in an increase in peak intensity and a change in the shape of the SPR peak. The main influence of the medium pH is its ability to change the electrical charges of the biomolecules present in the phytoextract, which could change their reducing and capping abilities. At higher pH levels, many phenolic functional groups are present, which aids in the conversion of Ag^+ ions into more Ag^0. Furthermore, no significant change occurs when the pH exceeds 8 [10,12-13]. Figure 2(c) shows the effect of extraction time on the SPR absorbance of AgNPs. When increasing the extraction time, the SPR absorbance increased.

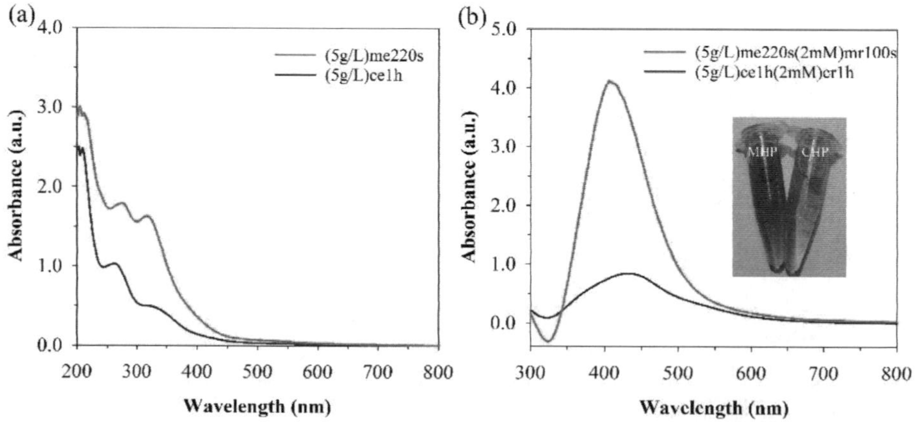

Figure 1. UV-visible absorption spectra of: (a) MLE from MHP and CHP and (b) the synthesized AgNPs from MHP and CHP. The insert is the corresponding sample.

Figure 2(d) shows the SPR absorbance of AgNPs at different microwave reaction times. It was observed that when the reaction time was increased from 60 s to 100 s, the SPR absorbance increased because of the increasing quantity and size of the AgNPs. When the time was greater than 100 s, the SPR absorbance decreased because the size, shape, and surrounding environment of the synthesized AgNPs underwent a change, indicating that the reaction ended within 100 s [9, 13-16].

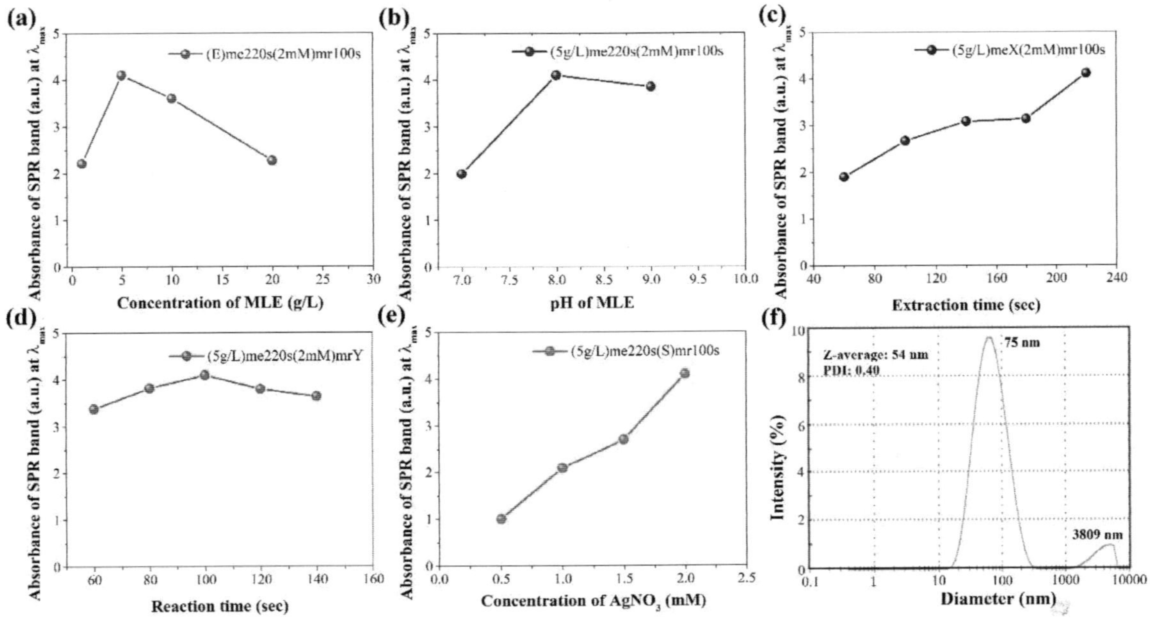

Figure 2. The absorbance of the SPR band of the synthesized AgNPs: (a) at different mint-to-water ratios (pH 8), (b) at different pH values, (c) at different microwave extraction times (pH 8), (d) at different microwave reaction times (pH 8), (e) at different $AgNO_3$ concentrations (pH 8), and (f) the DLS spectrum of the AgNPs synthesized by MHP.

Figure 2(e) shows the influence of $AgNO_3$ solution concentration on the formation of AgNPs. When the $AgNO_3$ solution concentration was increased from 0.5 mM to 2.0 mM, the SPR absorbance increased accordingly. This suggested that with more Ag^+ content, more AgNPs could be formed [17]. The highest SPR absorbance suggested that the number of AgNPs has reached its maximum, indicating the completion of the reaction [13-16]. The size distribution profile and the polydispersity index (PDI) of AgNPs synthesized by microwave heating were studied by the DLS technique, as shown in Figure 2(f). There were two peaks detected in the particle size distribution analysis, where peak 2 centered at 3809 nm represented the larger particle size of AgNPs. Peak 1, centered at 75 nm, represented the majority of its size. The higher intensity of peak 1 (94.1%) indicated the presence of a higher amount of smaller agglomerates of AgNPs compared with the lower intensity of peak 2 (5.9%), indicating a very small amount of larger agglomerates of AgNPs. Some agglomeration of AgNPs was also reported by other researchers [18]. Hence, the average particle size of AgNPs was 54 nm with a PDI of 0.40, indicating moderate polydispersity [14,19-20].

4. Conclusions

An eco-friendly and inexpensive plant-mediated, green method for the synthesis of AgNPs was successfully performed by using the leaf extract of mint as a stabilizing and reducing agent. The phytochemicals present in the extract could reduce the Ag^+ ions to Ag^0 and form AgNPs. Compared with the conventional heating process, the extraction and reaction times in the microwave-assisted method were 16 and 36 times faster, respectively, than those of the conventional heating method. The mint-to-water ratio, pH, extraction time, as well as the initial $AgNO_3$ concentration and microwave irradiation time, had obvious effects on the obtained AgNPs. This method produced stable, spherical AgNPs with an average particle diameter of 54 nm.

Acknowledgement
The authors would like to thank the Department of Materials Science and Engineering, Faculty of Engineering and Industrial Technology, Silpakorn University, and the Center of Excellence on Petrochemical and Materials Technology, Chulalongkorn University.

References

[1] Nouri A, Yaraki M-T, Lajevardi A, Rezaei Z, Ghorbanpour M and Tanzif M 2020 *Colloids Interface Sci. Commun.* **35** 100252.
[2] Khatoon A, Khan F, Ahmad N, Shaikh S, Rizvi S-M-D, Shakil S, Al-Qahtani M-H, Abuzenadah A-M, Tabrez S, Ahmed A-B-F, Alafnan A, Islam H, Iqbal D and Dutta R 2018 *Life Sci.* **209** 430.
[3] Li S, Al-Misned F, El-Serehy H and Yang L 2021 *Arab. J. Chem.* **14** 102931.
[4] Naowaboot J, Nanna U, Chularojmontri L, Songtavisin T, Tingpej P, Sattaponpan C, Jansom C and Wattanapitayakul S 2021 *Prev. Nutr. Food Sci.* **26** 157.
[5] Halake K and Lee J 2017 *J. Ind. Eng. Chem.* **54** 44.
[6] Kaeswurm J-A-H, Scharinger A, Teipel J and Buchweitz M 2021 *Molecules* **26** 4656.
[7] Choudhary N, Chatterjee M, Kumar S, Singh G and Suttee A 2021 *Mater. Today: Proc.* **45** 5362.
[8] Fatimah I 2016 *J. Adv. Res.* **7** 961.
[9] Peng H, Yang A and Xiong J 2013 *Carbohydr. Polym.* **91** 348.
[10] Philip D 2010 *Phys. E: Low-Dimens. Syst. Nanostructures* **42** 1417.
[11] Maity G-N, Maity P, Choudhuri I, Sahoo G-C, Maity N, Ghosh K, Bhattacharyya N, Dalai S and Mondal S 2020 *Int. J. Biol. Macromol.* **162** 1025.
[12] Ndikau M, Noah N-M, Andala D-M and Masika E 2017 *Int. J. Anal. Chem.* 2017 1.
[13] Carbone K, Angelis A-D, Mazzuca C, Santangelo E, Macchioni V, Cacciotti I, Petrella G, Cicero D-O and Micheli L 2020 *LWT - Food Sci. Technol.* **133** 110088.
[14] Banu R, Ramakrishna D, Reddy G-B, Veerabhadram G and Mangatayaru K-G 2021 *Mater. Today: Proc.* **43** 2265.
[15] Parit S-B, Karade V-C, Patil R-B, Pawar N-V, Dhavale R-P, Tawre M, Pardesi K, Jadhav U-U, Dawkar V-V, Tanpure R-S, Kim J-H, Jadhav J-P and Chougale A-D 2020 *Mater. Today Chem.* **17** 100285.
[16] Tamilarasi P and Meena P 2020 *Mater. Today: Proc.* **33** 2209.
[17] Kamal S-S-K, Vimala J, Sahoo P-K, Ghosal P and Durai L 2016 *Def. Sci. J.* **66** 368.
[18] Kumar R, Sharma P, Bamal A, Negi S and Chaudhary S 2017 *Green Process. Synth.* **6** 449.
[19] Kaplan O, Tosun N-G, Ozqur A, Tayhan S-E, Bilgin S, Turkekul I and Gokce I 2021 *J. Drug Deliv. Sci. Technol.* **64** 102641.
[20] Bhattacharjee S 2016 *J. Control. Release* **235** 337.

IOP Conf. Series: Materials Science and Engineering 1280 (2023) 012017 doi:10.1088/1757-899X/1280/1/012017

Effects of fiber surface modification on mechanical properties of short pineapple leaf fiber-carbon black reinforced natural rubber hybrid composites

S Thaiwattananon [1], S Thanawan [2] and T Amornsakchai [1,3,*]

[1] Department of Chemistry and Center of Excellence for Innovation in Chemistry, Faculty of Science, Mahidol University, Phuttamonthon 4 Road, Salaya, Phuttamonthon District, Nakhon Pathom, Thailand.
[2] Rubber Technology Research Center, Faculty of Science, Mahidol University, Nakhon Pathom 73170, Thailand.
[3] Center of Sustainable Energy and Green Materials, Faculty of Science, Mahidol University, Phuttamonthon 4 Road, Salaya, Phuttamonthon District, Nakhon Pathom, Thailand.

*Corresponding author: taweechai.amo@mahidol.ac.th

Abstract. At present, the world is focusing on the circular economy by turning waste into production resources again, reducing waste generation, and making it circulate in the system. This research focuses on using agricultural waste, which is pineapple leaves abundant in Thailand, to be processed into pineapple leaf fibers (PALF) and used to reinforce natural rubber. PALF has relatively high strength when compared to other natural fibers. On the other hand, the challenge of using PALF to reinforce natural rubber is the difference in the polarity of the fibers and rubber matrix. Thus, surface modification of PALF was studied to enhance the adhesion between the fiber surface and the matrix. Four systems were compared, i.e., untreated PALF, alkaline-treated, silane69-treated, and aminosilane-treated PALF. A unidirectional fiber-reinforced natural rubber composites were prepared with fixed fiber and carbon black contents of 10 and 30 phr, respectively. Mechanical properties of the composites were determined by tensile test and morphology was observed by scanning electron microscope (SEM). It was found that silane69-treated and aminosilane-treated PALF could reinforce the rubber matrix over the entire strain. However, silane69-treated and aminosilane-treated PALF had lesser reinforcement effect than alkaline-treated PALF at low strains. Despite a clear change in mechanical behavior, SEM could not tell the difference or changes that occur at the interface. Fiber surface modifications with silane69 and aminosilane have the potential to be further developed to improve properties of natural rubber composites and will, hopefully, increase the value of rubber products for Thailand industry.

1. Introduction

Extensive and excessive usage of resources causes it to decrease rapidly. One way to reduce resource usage is a circular economy. This concept is supported by the European Union and many national governments [1]. The concept of the circular economy has been defined in many ways [2]. One of the definitions is focused on the comprehensive use of resources and effective waste management. Thailand grows pineapples and exports a large number of canned pineapples each year causing tons of agricultural

IOP Conf. Series: Materials Science and Engineering 1280 (2023) 012017 doi:10.1088/1757-899X/1280/1/012017

waste which is pineapple leaf [3]. In addition, pineapple leaf fiber (PALF) has high strength compared to other natural fibers [4]. Therefore, the use of PALF has received a great deal of interest in terms of strength and waste management. Numerous studies have reported the performance of PALF-reinforced rubber composites. The two main areas being studied are rubber matrix modification and fiber surface treatment. For example, varying crosslink density and carbon black loading, rubber mastication time, carbon black distribution, and rubber polarity improvement, can enhance the fiber-matrix compatibility and modify the stress-strain curves [5-9].

This work focused on increasing the compatibility between the fiber surface and rubber matrix by chemically modifying the surface of PALF, i.e., alkaline treatment, silane69, and aminosilane modifications. After surface modification, the modified fibers were used to reinforce natural rubber with a fixed amount of carbon black at 30 phr. The samples were tested by a tensile testing machine, chemical surface modification was confirmed with attenuated total reflectance Fourier transform infrared spectroscopy (ATR-FTIR) and the fracture surfaces were investigated with a scanning electron microscope (SEM).

2. Experimental

2.1 Materials

Pineapple leaves were collected from a plantation in Bang Yang district, Phitsanulok province, Thailand. The fresh leaves were processed to separate pineapple leaf fiber (PALF) and non-fibrous material (NFM) by the proposed method by Kengkhetkit and Amornsakchai [10]. Natural rubber (NR) was Standard Thai Rubber grade STR5L purchased from MBJ Enterprise Co. Ltd. Thailand. Carbon black (CB) was N330 grade manufactured by Thai Carbon Black Public Co., Ltd. Bis-[γ-(triethoxysilyl)-propyl]-tetrasulfide (silane69) and (3-Aminopropyl) triethoxysilane (Aminosilane) were purchased from Sigma-Aldrich. The adhesion promoter was carboxylated liquid polyisoprene (LIR-410), which was donated by Kuraray Co., Ltd., Japan. The other chemicals and curing agents were commercial grades.

2.2 Fiber surface modification

Pineapple leaf fiber (PALF) was treated with an alkaline solution (10 % w/v NaOH) for 30 min then washed with distilled water until the pH is neutral, called alkaline treated pineapple leaf fiber (APALF). After that, APALF was further treated with silane69 and aminosilane following the method reported by Surajarusarn and Amornsakchai [11].

Table 1. The sequence of mixing and rubber compound formulation.

Step	Ingredients	Time (min)
1	NR (STR5L)	0-2
2	Masterbatch of fiber	2-5
3	Masterbatch of carbon black	5-8
4	Zinc oxide and stearic acid	8-10
5	Sulfur and N-cyclohexyl Benzothiazole Sulfenamide	10-13
Total time		13

2.3 Rubber composite preparation

Masterbatch of carbon black (MBCB) was prepared with laboratory internal mixer with a fill factor of 0.75, and the weight ratio of rubber to carbon black was 60:40. Masterbatch of fiber was prepared by using a laboratory two-roll mill. NR and MBCB were masticated for 16 min and then, the fiber was gradually added while mixing for 12 min. The sequence of mixing and rubber compound formulation is shown in Table 1. The amount of the mixture was fixed at 40 g and the total mixing time is 13 min. The rubber compound was mixed and sheeted in the same direction along with the fiber orientation. Then, the compound was cut to fit a square mold with a thickness of 1 mm. The compression machine was set at the temperature of 150 °C and the pressure was 10.3 MPa.

3. Results and discussion

ATR-FTIR spectrums of modified fibers are shown in Figure 1. The intensity of the peak at wavenumber of 3200 – 3400 cm^{-1} (OH stretch) decreased due to the substitution after the treatment. Moreover, some of the peaks at the wavenumber of 1460 – 1560 cm^{-1} and 2850 – 2920 cm^{-1} change in appearance indicating the change of functional groups after chemical treatment.

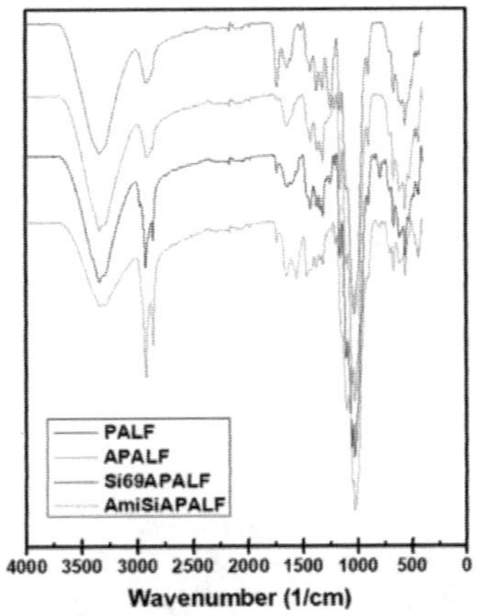

Figure 1. ATR-FTIR spectrum of four different types of PALF.

Stress-strain curves of the composite are shown in Figure 2. With the addition of PALF, the graph changes significantly in the low-strain regions similar to that reported in the literature [12]. There is a significant increase in stress causing the modulus to be significantly higher. After the yield-like point, the stress slightly decreased due to the debonding of fibers. When reaches 200% strain, the stress starts to increase again due to the strain-induced crystalline of natural rubber. The curves of APALF-reinforced composites are similar to that of PALF with a decrease in stress after the yield-like point, but the difference is that the stress at the yield-like point is much higher, and the stress upturn is approximately 230% strain. After sodium hydroxide treatment, lignin and hemicellulose were removed [13] and this provides better friction and higher stress after the yield-like point. Both PALF and APALF composites had similar elongation at break of about 370% strain. The stress-strain curves of silane69 and aminosilane-treated PALF are very similar to that of PALF composite in the early part but display a gradual rise in stress after the yield-like point. The stress upturn of silane69-treated PALF composites occurs at lower strain and stress is higher than other composites over the whole range. This change occurs possibly because of hydrogen and chemical bonding between the modified fibers and carboxylated group of the adhesion promotor [14].

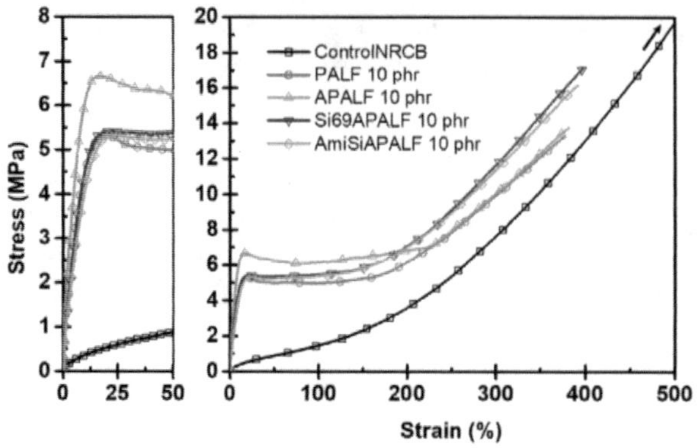

Figure 2. Stress-strain curves of natural rubber composites containing 10 phr of various types of fibers and 30 phr of carbon black.

The tensile fracture surfaces of all composites are shown in Figure 3. Failure occurs with a lot of fibers being pulled out. Both single fibers and fiber bundles can be observed in all four samples. Not much difference between the four composite samples can be seen.

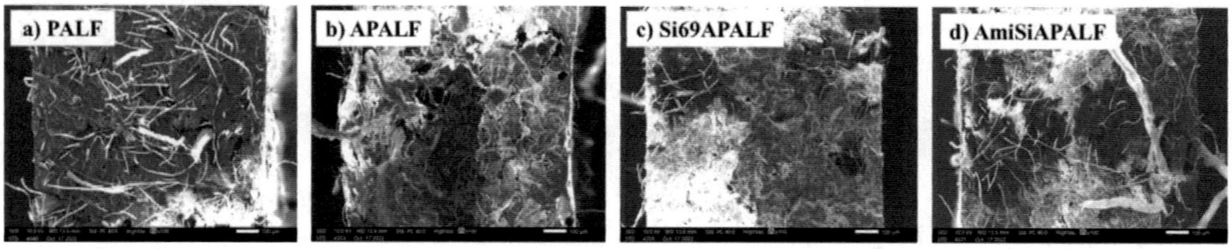

Figure 3. SEM micrographs of tensile fractured specimens of (a) NR-PALF, (b) NR-APALF, (c) NR-Si69APALF, (d) NR-AmiSiAPALF.

4. Conclusions

Surface modification of fiber has played an important role to improve adhesion between the fiber surface and the rubber matrix. Adding a small amount of PALF causes a significant change in the stress-strain curve. Alkaline-treated PALF appears to be the best way to enhance the modulus of rubber composite in low-strain regions. However, the stress was dropped after the yield-like point, and the effect of fiber reinforcement was slightly reduced. For silane69 and aminosilane-treated PALF, no drop in stress over the entire strain is seen. Moreover, silane69 and aminosilane-treated PALF also increase elongation at break of the composites. Modification of the PALF surface has an efficiency to improve interfacial adhesion between the fiber surface and rubber matrix. The reinforcement efficiency of PALF in natural rubber should expand the scope of the application of rubber composites, providing a greater opportunity to increase industrial value.

Acknowledgement

This work was supported by Department of Chemistry and The Center of Excellence for Innovation in Chemistry (PERCH-CIC), Faculty of Science, Mahidol University, and Grant to Support Graduate Students in Academic Presentations in Thailand, Faculty of Graduate of Studies, Mahidol University.

References

[1] Korhonen J, Honkasalo A and Seppälä J 2018 *Ecol econ* **143** 37.
[2] Kirchherr J, Reike D and Hekkert M 2017 *Resour Conserv Recycl* **127** 221.

[3] Wattanakul T, Nonthapot S and Watchalaanun T 2021 *Australas. Account. Bus. Finance J.* **15** 119.
[4] Kalia S, Kaith B S and Kaur I 2009 *Polym. Eng. Sci.* **49** 1253.
[5] Pittayavinai P, Thanawan S and Amornsakchai T 2016 *Polym. Test.* **54** 84.
[6] Yantaboot K and Amornsakchai T 2017 *Polym. Test.* **57** 31.
[7] Yantaboot K and Amornsakchai T 2017 *Polym. Test.* **61**. 223.
[8] Hariwongsanupab N, Thanawan S, Amornsakchai T, Vallat M F and Mougin K 2017 *Polym. Test.* **57** 94.
[9] Surajarusarn B, Hariwongsanupab N, Schrodj G, Garreau S, Mougin K and Amornsakchai T 2019 In *Euro-Mediterranean Conference for Environmental Integration,* Springer, Cham, pp. 441-446.
[10] Kengkhetkit N and Amornsakchai T 2012 *Ind. Crops Prod.* **40** 55.
[11] Surajarusarn B, Hajjar-Garreau S, Schrodj G, Mougin K and Amornsakchai T 2020 *Polym. Test.* **82** 106289.
[12] Prukkaewkanjana K, Thanawan S and Amornsakchai T 2015 *Polym. Test.* **45** 76.
[13] Surajarusarn B, Traiperm P and Amornsakchai T 2019. *Sains Malays.* **48** 145.
[14] Surajarusarn B, Thaiwattananon S, Thanawan S, Mougin K and Amornsakchai T 2021 *Fibers Polym.* **22** 2543.

Mechanical properties of the laser powder deposition and laser powder bed fusion printed 316L

M Hietala[1*], T Rautio[1], J Mäkikangas[1] and A Järvenpää[1]

[1] University of Oulu, Kerttu Saalasti Institute, FI-85500 Pajatie 5 Nivala, Finland

*Corresponding author: mikko.hietala@oulu.fi

Abstract. Metal 3D printing technologies have made it possible to produce different parts, but the mechanical properties of the parts connected using different 3D printing technologies have not yet been sufficiently studied. In the study, 316L stainless steel was laser powder deposited (DED) with laser cladding head on the laser powder bed fusion (PBF-LB) printed 316L and the joint was investigated experimentally. The microstructure of the joint was examined from the cross-sections of the joint by optical microscopy. The mechanical properties of the joint were evaluated by hardness measurements and tensile strength tests. The microstructure of the deposited structure was similar to the typical 316L printed structure, but at the joint, due to the higher heat input of the DED process, an increase in grain size was clearly visible. Based on the hardness measurements, the hardness profile of the joint was very even, and the average hardness was approximately 305 HV. The tensile strength of the PBF-LB and DED joint was 549 MPa which is approximately 11% lower than the tensile strength of PBF-LB printed 316L.

1. Introduction

The production of various metal parts using metal 3D printing methods has become very common recently. As a printing method, the laser powder bed fusion laser (PBF-LB) method is very popular for printing small, precision-demanding, and diverse pieces [1,2]. However, PBF-LB is not suitable for manufacturing very large parts. Direct energy deposition (DED) printing methods are better for manufacturing larger parts. However, it is possible to combine parts made with two different printing methods by printing on top of the part [3].

So far, very little research has been done on the mechanical properties of the joint of parts made with two different printing methods. From the industry's point of view, more investment must be made in the research of hybrid joints in the future.

Kim *et al.* [4] investigated interfacial characteristics and crack propagation of 630 stainless steel manufactured by DED on PBF substrate. Their observation was that the DED part showed a lower hardness than the PBF part. Yan *et al.* [5] studied microstructure and mechanical properties of austenitic stainless steels manufactured by DED. They observed that 316L manufactured by DED had a greater mechanical property than 316L sheet metal.

Yazar *et al.* [6] investigated effect of DED process parameters on microstructure of 316L. They observed a columnar to equiaxed transition of the grains in deposited material. Zhou *et al.* [7] studied DED of

316L with coaxial wire-powder feeding. They found that microstructure mainly consist of columnar grains that grow epitaxially across the deposited layers.
The aim of this study is to find out the mechanical properties of the PBF-LB and DED 316L joint.

2. Experimental methods

2.1 Printing equipment

The PBF part was manufactured utilizing an SLM 280 HL machine (manufacturer: SLM Solutions Group AG) with a 280 x 280 mm platform. The PBF part was printed with following parameters: laser power (P) of 200 W, speed (v) of 800 mm/s, hatch spacing (h) of 120 µm, layer thickness (t) of 30 µm and laser spot diameter 0.1 mm. The 316L powder used in the PBF printing was supplied by Carpenter Additive of Carpenter Technology Corporation and the chemical composition of the powder is presented in Table 1.

The DED part was printed utilizing Trumpf HLD-4002 disc-laser and Precitec YW-52 coaxial cladding head. The DED part was printed with following parameters: laser power (P) 2900 W, speed (v) 15 mm/s, focal point -8 mm, tilt angle 5°, nozzle distance from the part 12.5 mm, powder feed 15% (25.5 g/min) and feed gas argon (Ar) pressure 3 bar 4.5 l/min. The powder used in DED printing was 316L Metco 41C supplied by Sulzer Ltd.

Table 1. The chemical composition % of the powder.

Powder	Cr	Ni	Mn	Mo	C	Si	Cu	N	O
PBF	17.6	12.5	0.66	2.38	0.02	0.65	0.02	0.09	0.03
DED	17	12	-	2.5	0.03	1	-	-	-

2.2. Mechanical characterization

Tensile tests were conducted utilizing a universal material testing machine (Instron 8802) and test were made according to the standard SFS-EN ISO 68921:2016. Test were conducted with a constant loading rate of 1.0 mm/min. Microhardness measurements were conducted with an Innovatest Falcon 511 using Vickers indenter with 0.2 kg load. In the hardness measurements, the distance between the measuring points was 0.1 mm. Specimens for microstructure evaluation were polished and etched. Microstructure was analysed with optical microscope Keyence VHX 2000.

3. Results and discussion

3.1 Microstructure

The microstructure of the PBF-LB printed 316L is shown in Fig. 1a. The microstructure of PBF-LB clearly shows the boundaries of the melt pool, which have formed between the printing layers. The boundaries of the melt pool form a microstructure resembling fish scale pattern. Fig. 1b shows the substructure created during PBF-LB printing.

Fig. 1c shows the interface of the two printed structures. Heat from DED printed layers caused epitaxial grain growth in the PBF-LB microstructure, leading to a coarse columnar grain structure which can be seen in Fig. 1c. Also, in the microstructure of the DED printed 316L, oriented grains can be seen at the interface of the parts.

Fig. 1d shows the microstructure of the DED printed part, which is austenitic with a small amount of ferrite. The heat input of the DED process was so great that delta-ferrite can be found in the microstructure.

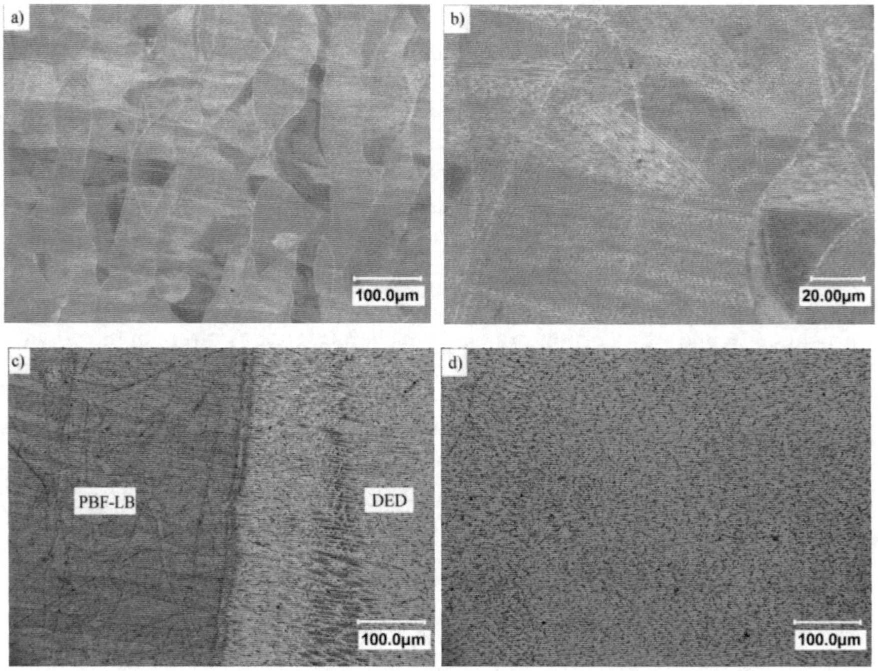

Figure 1. Microstructure of (a) PBF-LB 316L; (b) PBF-LB 316L substructure; (c) interface of PBF-LB and DED printed 316L and (d) DED 316L.

3.2 Microhardness

The results of the hardness measurements can be seen in Fig. 2. Based on the hardness measurements, the base material hardness of both PBF-LB and DED printed 316L was almost the same. A slightly softer area can be seen at the interface of the printed parts, which is due to a large heat input to the area. The softer area is explained by the larger grain size observed in the microstructure of the area.

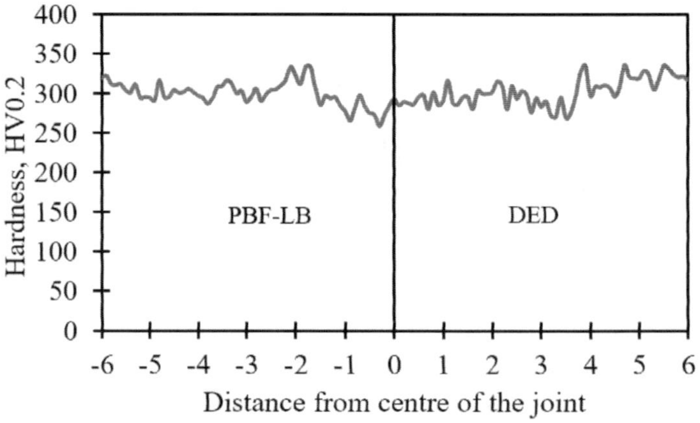

Figure 2. Hardness of the interface of PBF-LB and DED printed parts.

3.3 Tensile strength of the joint

Tensile tests were performed on the PBF-LB and DED printed interface in order to find out the mechanical properties of the joint. The results were compared to the strength of PBF-LB printed 316L. Fig. 3 shows that the yield strength of the PBF-LB and DED joint was approximately 434 MPa and ultimate tensile strength 549 MPa. The yield strength of PBF-LB 316L was approximately 488 MPa, which is about 12.5% higher than the yield strength of the PBF-LB and DED joint. However, the elongation of the PBF-LB and DED joint was approximately 28% bigger than the elongation of PBF-LB 316L. The grain growth seen in the microstructure of the PBF-LB and DED interface clearly affected the strength of the PBF-LB and DED joint.

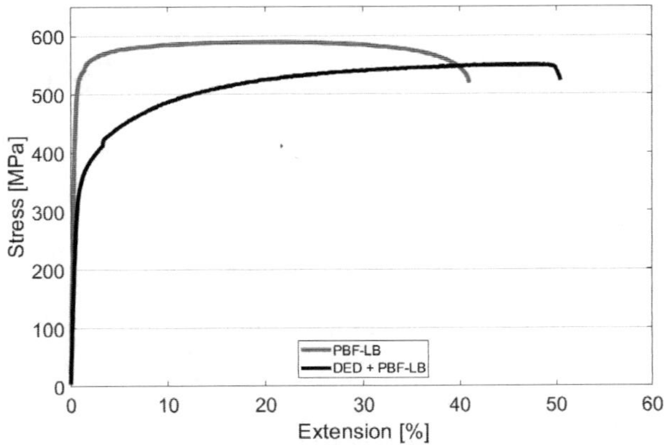

Figure 3. Tensile strength of PBF-LB 316L and interface of PBF-LB and DED printed part.

4. Conclusions

In this paper, 316L stainless steel was laser powder deposited (DED) on the laser powder bed fusion (PBF-LB) printed 316L and the joint was investigated experimentally. The following conclusions can be drawn:

- The heat from DED printed layers caused epitaxial grain growth in the PBF-LB 316L microstructure, leading to a coarse columnar grain structure in the interface of the parts. In the microstructure of the DED printed 316L, oriented grains can be seen at the interface of the parts.
- A slightly softer area can be seen at the interface of the printed parts, which is due to a larger grain at the interface.
- The yield strength of PBF-LB and DED 316L joint was approximately 434 MPa.

Acknowledgements

The authors would like to acknowledge the financial support received from the Council of Oulu Region and the European Union (European Regional Development Fund) for the M3D project and Business Finland for the DREAMS project. The authors would like to acknowledge the financial support received from City of Nivala, Nivala industrial park Ltd, NIHAK Nivala-Haapajarvi region registered association.

References

[1] Horstkotte R, Heinrich F, Prümmer M, Arntz K and Bergs T 2021 *Procedia CIRP* **96** 97.
[2] Köhnen P, Letang M, Voshage M, Schleifenbaum J H and Haase C 2019 *Addit. Manuf.* **30** 100914.
[3] Yusuf S M, Zhao X, Yang S and Gao N 2020 *Mater. Lett.* **284** 128928.
[4] Kim T G, Shin G Y and Shim D S 2022 *Mater. Sci. Eng. A.* **835** 142657.
[5] Yan Z, Zou K, Cheng M, Zhou Z and Song L 2021 *J. Mater. Res. Technol.* **15** 582.
[6] Yazar K U, Pawar S, Park K S and Choi S H 2022 *Mater. Charact.* **191** 112148.
[7] Zhou Y and Ning F 2022 *Manuf. Lett.* **33** 686.

IOP Conf. Series: Materials Science and Engineering 1280 (2023) 012019 doi:10.1088/1757-899X/1280/1/012019

Fabrication and Properties of Gelatin/Dialdehyde Chitosan Films

S Hemsri[1,*], N Junwattarunggu[1], W Rueangsawang[1] and S Aramsrithum[1]

[1] Department of Materials Science and Engineering, Faculty of Engineering and Industrial Technology, Silpakorn University, Sanamchandra Palace Campus, Nakhon Pathom 73000, Thailand.

*Corresponding author: sudsirihemsri@yahoo.com

Abstract. Non-biodegradable petroleum-based packaging film waste has led to a severe environmental pollution and human health problems. Gelatin and chitosan are attractive biopolymers for biomedical and food packaging applications due to biocompatibility, biodegradability, renewability, nontoxicity, and good gas barrier properties. However, disadvantages of the gelatin and chitosan are their low physicochemical stability and poor mechanical properties. In this study, dialdehyde chitosan was synthesized by periodate oxidation to use as a cross-linking agent for gelatin. The effect of the incorporation of the dialdehyde chitosan with 10, 15 and 20 wt% on mechanical and swelling properties as well as wettability of the gelatin films was investigated. The results reveal that the mechanical strength and swelling properties of the gelatin films were improved as an increase in the dialdehyde chitosan content, whereas the addition of the dialdehyde chitosan led to a decrease in water swelling ability in the gelatin films. Moreover, the wettability of the gelatin/dialdehyde chitosan films showed hydrophobic behaviour. Thus, the obtained gelatin/dialdehyde chitosan films as green polymeric products have a potential for food packaging fields.

1. Introduction

Gelatin (G) and chitosan (CS) are attractive natural polymers for biomedical and food packaging due to many significant properties such as biocompatibility, biodegradability, renewability, nontoxicity, biological activity and good gas barrier properties. Gelatin, a hydrocolloid protein, is extracted from collagen, while chitosan, a polysaccharide derived from the partial deacetylation of chitin, is a cationic linear biopolymer. However, disadvantages of the gelatin and chitosan are their low physicochemical stability (high swelling and solubility), high biodegradation rate and poor mechanical properties, leading to a limiting their use in many applications [1-2]. Thus, it is necessary to crosslink them to enhance their physicochemical stability and also mechanical strength in wet or aqueous environment [1-2]. The chemical crosslinking by using crosslinking agents is one of effective methods to improve mechanical properties and water stability of proteins and polysaccharides. However, many effective crosslinking agents for proteins such as glutaraldehyde, glyoxal and formaldehyde are toxic [3-5]. Nowadays, new crosslinking agents, for example natural macromolecular dialdehydes are promising compounds due to their nontoxicity, biodegradability and high reactivity, especially with amino groups in protein chains [1,6]. There are many studies concerning the use of dialdehyde carboxymethyl cellulose and dialdehyde quince seed gum as crosslinking agents for gelatin [7-10]. Moreover, dialdehyde starch and dialdehyde cellulose nanocrystal have been reported as crosslinking agents for gelatin/chitosan scaffolds and gelatin/chitosan films [1,11]. However, to the best of our knowledge in the available literature, no work has been studied related to gelatin crosslinked with dialdehyde chitosan. Thus, the aims of this study were to synthesize a dialdehyde chitosan (DAC) via periodate oxidation to use as a cross-linking agent for gelatin films and to study the effect of DAC content on mechanical properties, swelling ability and water wettability of the gelatin films. We expect that the prepared gelatin crosslinked with dialdehyde chitosan (G/DAC) films are promising green polymeric products for food packaging application.

2. Experimental

2.1. Materials

A commercial gelatin (G) derived from pork-skin collagen with a 250 bloom degree was purchased from Nitta Gelatin Inc., Thailand. A commercial chitosan (CS) with grade of SSA190/301CF, a molecular mass of 250-2100 kDa was supplied from Marine Bio Resources Co. Ltd., Thailand. Sodium periodate was obtained from Sigma-Aldrich.

2.2. Preparation and characterization of dialdehyde chitosan (DAC)

Chitosan was dissolved in 1 %v/v of acetic acid solution for 2 h and sodium periodate was dissolved in distilled water. Dialdehyde chitosan was synthesized by oxidation reaction of chitosan with sodium periodate aqueous solution at 40°C in the dark for 3 h, according to the DAC synthesis method in the literature [5]. The weight ratio of chitosan and sodium periodate was at 1:1.

The morphologies of the chitosan (CS) and the obtained dialdehyde chitosan (DAC) were observed with a Field Emission Scanning Electron Microscope (TESCAN MIRA3 FEG-SEM, Czech Republic) operated at an accelerated potential of 10 kV. The powder samples were sputter-coated with gold. The Attenuated Total Reflectance Fourier Transform Infrared (ATR-FTIR) analysis was used to determine the chemical structure of the chitosan (CS) and dialdehyde chitosan (DAC). The CS and DCS spectra were recorded using Perkin Elmer Frontier FTIR with ATR-FTIR mode within a spectral range of 4000-400 cm^{-1}.

2.3. Preparation and testing methods of gelatin (G) and gelatin/dialdehyde chitosan (G/DAS) films

Gelatin/Dialdehyde chitosan (G/DAC) films were prepared by dissolving 4 g of gelatin in 130 ml of distilled water, stirring the gelatin solution at room temperature for 30 min and at 45°C. respectively. Then dialdehyde chitosan was added at 10, 15 and 20 wt% of gelatin into the gelatin solution and the mixture was stirred until completely dissolved. Glycerol at 20 wt% of solid content was added into the G/DAC solution and stirred for 2 h. After that, the solution was poured into the teflon mold and heated in an oven at 50°C for 24 h for crosslinking and drying the films.

Tensile properties of all films were performed using Universal Testing Machine (Instron 5969, Instron Engineering corporation), according to ASTM D882. Samples are conditioned at relative humidity of 50% at room temperature prior to the testing.

Water swelling ability was determined by the gravimetric method. All G and G/DAS films were cut into a shape with 1 x 1 cm^2. The dry film was weighted (W_1) before testing. The film was then soaked in deionized water at room temperature for 60 min. After the water soaking, the residue water on the film surface was removed by a tissue towel paper and then the wet film was weighted (W_2). The swelling ratios of each sample were calculated using equation (1):

IOP Conf. Series: Materials Science and Engineering 1280 (2023) 012019 doi:10.1088/1757-899X/1280/1/012019

$$\text{Swelling ratio} = \frac{W2 - W1}{W1} \times 100\%$$

Static water contact angle measurement on surface of G and G/DAC films was conducted using an optical contact angle meter with at least 3 measurements in each sample.

3. Results and discussion

3.1. Characterization of dialdehyde chitosan (DAC)

The morphologies of pure chitosan (CS) and the synthesized dialdehyde chitosan (DCS) are presented by SEM images as shown in Figure 1. In the pure CS image (Fig.1a, at 100x magnification), flake-like structures with varying sizes was observed. After the periodate oxidation of CS, morphology of round shape DAC particles became smaller with mean size around 1-5 μm, as seen in Fig.1b (with the 5000x magnitude). Moreover, some small needle-shape particles were observed on DAC surface. This morphology observation of DAC in our research is similar to the work of Wegrzynowska-Drzymalska *et al.* [5] and Grabska-Zielińska *et al.* [12].

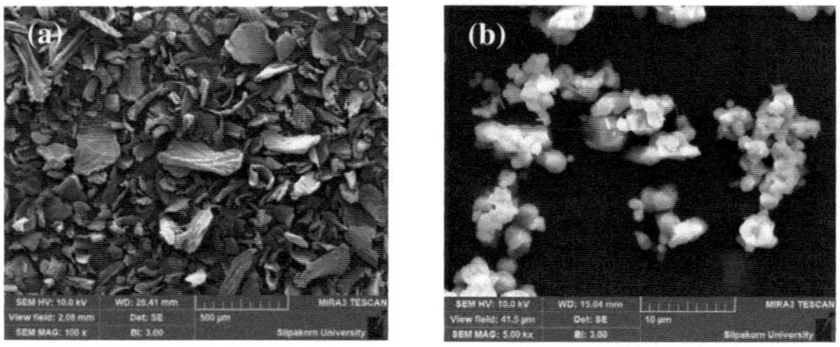

Figure 1. SEM images of (a) chitosan (CS) with 100x magnitude and (b) dialdehyde chitosan (DAC) with 5000x magnitude.

The ATR-FTIR analysis was carried out in order to confirm chemical structure of the synthesized dialdehyde chitosan (DCS) compared with pure chitosan (CS). Figure 2 illustrates the ATR-FTIR spectra of CS and DCS as well as the schematic representation of periodate oxidation reaction of chitosan (CS).

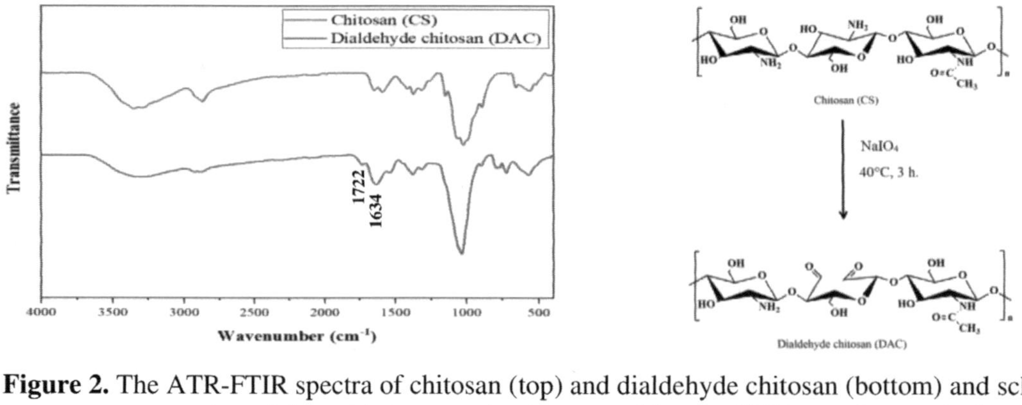

Figure 2. The ATR-FTIR spectra of chitosan (top) and dialdehyde chitosan (bottom) and schematic representation of the oxidation reaction between chitosan (CS) by sodium periodate.

In spectrum CS, broad peak at 3356 and 3295 cm^{-1} attributed to O-H and N-H stretching was observed. Furthermore, C-H stretching, C=O stretching of amide I, N–H bending of amide II and C–O stretching appeared at 2875, 1651, 1589 and 1023 cm^{-1}, respectively [5]. Considering the spectrum of dialdehyde chitosan, new adsorption peaks at 1722 and 1634 cm^{-1} are attributed to C=O stretching vibration of aldehyde groups, resulting from the periodate mediated oxidation of chitosan [13-14]. Moreover, the changes in the shape and intensity of the O-H stretching of hydroxyl groups at 3307 cm^{-1} clearly observed due to the opening of cyclic structure of chitosan saccharide units after the periodate oxidation [13].

3.2. Mechanical properties of gelatin and gelatin/dialdehyde chitosan films

Mechanical properties including Young's modulus, maximum tensile strength and % elongation at break of the gelatin and G/DAS films with various content of DAC crosslinking agent at 10, 15 and 20 wt% were exhibited in Figure 3. As the results, the gelatin crosslinked with DAC (G/DAC) films revealed higher values of Young's modulus and maximum tensile strength compared to the pure gelatin film. Furthermore, the modulus and tensile strength of the G/DAC films increased as an increase in DAC content. It can be explained by the presence of the reaction of aldehyde groups in DAC and amine groups of gelatin through the Schiff base reaction, leading to crosslink reaction between gelatin chains [13]. Thus, this result caused an enhancement in modulus and maximum tensile strength of the gelatin film added with DAC.

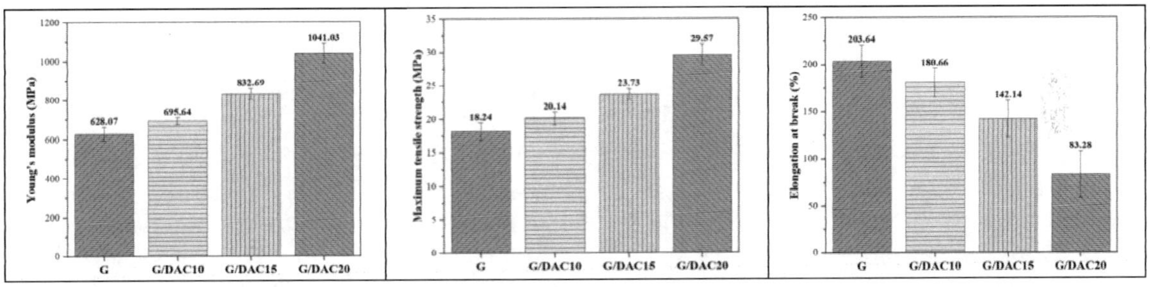

Figure 3. Young's modulus, maximum tensile strength and elongation at break of pure gelatin (G) and gelatin cross-linked by 10%, 15% and 20% of dialdehyde chitosan (DAC) films.

3.3. Swelling ability and static water contact angle of gelatin and gelatin/dialdehyde chitosan films

The swelling ratios were calculated to evaluate the water stability of G and G/DAC films as shown in Table 1. The swelling ratio of uncrosslinked gelatin film (G) could not be obtained because the swollen film was broken up into small pieces after 5 min of immersion in water, indicating the high hydrophilic behaviour of gelatin film. As shown in Table 1, with an increasing DAC concentration, the swelling ratio of the G/DAC films decreased. This is due to a presence of a more compact network structure owning to crosslink formation in the G/DAC films, thereby preventing the films from swelling in water.

Table 1. The swelling ratios and contact angles of gelatin and gelatin crosslinked dialdehyde chitosan films

Sample	Swelling ratio (%) at 60 min	Contact angle (degree)
G	-	113.93±0.78
G/DAC10	784.66	111.65±0.70
G/DAC15	724.05	102.73±1.29
G/DAC20	449.36	98.17±0.98

The contact angle measurement of the gelatin (G) and gelatin crosslinked with dialdehyde chitosan (G/DAC) films was conducted in order to evaluate hydrophobicity and hydrophilicity of film surface. As can be seen in Table 1, the contact angle of G and G/DAC films were more than 90°, indicating that surfaces of all films were hydrophobic. Interestingly, the uncrosslinked gelatin film provided high contact angle comparing to all G/DAC films. In addition, an increase in DAC concentration led to a decrease in contact angle. This may be because of more polarity of aldehyde groups in DAC chains, thus the G/DAC films became more hydrophilic than the G films.

4. Conclusions

In this study, the dialdehyde chitosan (DAC) was successfully synthesized via the periodate oxidation to use as a cross-linking agent for gelatin. The chemical structure of the obtained DAC was confirmed by ATR-FTIR spectroscopy. The G/DAC films provided a superior mechanical properties and swelling ability to the uncrosslinked gelatin film. An increasing content of DAC crosslinking agent in the gelatin resulted in an increase in modulus, tensile strength and water contact angel of the gelatin films, whereas elongation at break and water swelling property decreased. As the result of the water wettability test of film surface, although the contact angles of the G/CDAC films were lower than those the pure gelatin film, the surface of G/DAC films showed hydrophobic characteristic. Based on the results obtained, the G/DAS biofilms can potentially be used in food packaging application. However, further studies such as water vapor and oxygen permeability as well as antimicrobial ability will be investigated.

Acknowledgement

The authors would like to acknowledge the financial support and testing instruments to Department of Materials Science and Engineering, Faculty of Engineering and Industrial Technology, Silpakorn University.

References

[1] Kaczmarek B, Sionkowska A, Monteiro FJ, Carvalho A, Łukowicz K and Osyczka AM 2018 *Biomed. Mater.* **13** 015016.
[2] Reddy MSB, Ponnamma D, Choudhary R and Sadasivuni KK 2021 *Polymers* **13** 1105.
[3] Rhim J-W, Gennadios A, Weller CL, Cezeirat C and Hanna MA 1998 Ind. Crops Prod. **8** 195.
[4] Münster L, Capáková Z, Fišera M, Kuˇritka I and Vícha L 2019 *Carbohydr. Polym.* **218** 333.
[5] Wegrzynowska-Drzymalska K, Grebicka P, Mlynarczyk DT, Chelminiak-Dudkiewicz D, Kaczmarek H, Goslinski T and Ziegler-Borowska M 2020 *Materials* **13** 3413.
[6] Dou Y, Zhang L, Zhang B, He M, Shi W, Yang S, Cui Y and Yin G 2020 *Polymers* **12** 158.
[7] Mu C, Guo J, Li X, Lin W and Li D 2012 *Food Hydrocoll.* **27** 22.
[8] Guo J, Li X, Mu C, Zhang H, Qin P and Li D 2013 *Food Hydrocoll.* **33** 273.
[9] Sethi S, Kaith BS, Kaur M, Sharma N and Khullar S 2020 *J. Chem. Sci.* **132** 15.
[10] Maroufi LY, Shahabi N, Ghanbarzadeh MD and Ghorbani M 2022 *Food Bioproc. Tech.* **15** 693.
[11] Wegrzynowska-Drzymalska K, Mlynarczyk DT, Chelminiak-Dudkiewicz D, Kaczmarek H, Goslinski T and Ziegler-Borowska M 2022 *Int. J. Mol. Sci.* **23** 9700.
[12] Grabska-Zielińska S, Sionkowska A, Olewnik-Kruszkowska E, Reczynska K and Pamuła E 2021 *Int. J. Mol. Sci.* **22** 3391.
[13] Lv Y, Long Z, Song C, Dai L, He H and Wang P 2013 *BioResources* **8** 6161.
[14] Wu M, Lv W, Wang F, Long Z, Chen J and Dong C 2018 *Cellulose Chem. Technol.* **52** 43.

IOP Conf. Series: Materials Science and Engineering 1280 (2023) 012020 doi:10.1088/1757-899X/1280/1/012020

Effect of Blowing Agents on Properties of Cellular rubber of Natural rubber / Chloroprene blends

C Thongpin[1,*] and L Angsupanich[1]

[1]Department of Materials Science and Engineering, Faculty of Engineering and Industrial Technology, Silpakorn Univerisity, Sanamchandra Palace Campus, Nakhonpathom, 73000, Thailand.

*Corresponding author: Thongpin_c@su.ac.th, cmaterials@hotmail.com

Abstract. Cellular or sponge rubber can be by expansion of fluid state by gas produced from decomposition of foaming agents [1]. It can be manufactured from various types of rubber and chemicals depending on properties required [2]. Natural rubber can give excellent mechanical properties except for thermal ageing [3]. The objectives of this work were to improve thermal ageing of NR foam by incorporating chloroprene rubber, CR, into NR and foaming NR/CR blends. In order to study the effects of the type of chemical blowing agents on the properties of cellular NR/CR rubber blends, azodicarbonamide (ADC) and oxybisbenzenesulfonyl hydrazide (OBSH) were used as blowing agents and compared. The cure characteristic, mechanical, morphological, and flammability properties of the foams were investigated. The results indicated an increasing chloroprene rubber loading caused the increase in cure time, but decreased in cure rate index. In both cases of foaming agents, the cellular rubbers were found that tensile and tear strength were decreased with CR rubber loading, but the secant modulus and hardness were increased. Compression set was decreased with CR rubber loading. It is also showed that the flammability of NR/CR foam with high CR contents (more than 80%) was better than that with low CR contents. It is also found that using OBSH as foaming agent can produce cellular rubber containing better cell density and cell size distribution than ADC.

1. Introduction

Foam and sponge rubber are a class of cellular rubber which advantages in its light weight while its strength is not sacrificed. The preparation of foam can proceed by expanding the rubber in a fluid state to a low-density cellular state and then preserving this state. The expansion of fluid state is by gas produced from decomposition of foaming agents [1]. The properties required is mainly designed by rubber types used and additives incorporated [2]. Rubber foam has various applications such as mattresses, pillows, dolls, cushions, flooring, and upholstery foam. Nonetheless thermal ageing and weather resistance are some weaknesses of products derived from NR [3]. There are many researches reporting that blending of NR with other elastomer or rubber can give rise to improvement in its properties for some extent [4]. Chloroprene (CR) or Neoprene has possessed not only a good feature in weather and flame proof but also in oil and fuel resistance. Additionally, CR could give excellent mechanical performance and wear resistance as well as NR. Even so, CR has some drawbacks over NR in terms of resistance to strong acids.

IOP Conf. Series: Materials Science and Engineering 1280 (2023) 012020 doi:10.1088/1757-899X/1280/1/012020

This research is aimed to improve thermal ageing flame retardant of the foam in term of LOI and UL-94 class of flammability test. However, cell morphology of rubber foam could be controlled by the action of blowing agent thus the second objective is to study the effect of blowing agents used for blowing rubber on foam morphology and flammability. In this work NR/CR blend is compatibilized using MA-g-IR in all compositions.

2. Experimental

2.1 Preparation of NR/CR blends: Blending NR with CR was carried out on a Two Roll Mill (LabTech. Co. Ltd., Thailand) by firstly masticating NR (STR5L, J.P. Rubber Co.Ltd. Thailand) for 30 minutes after that various contents of CR (M-30, Innovation Tech. Co.Ltd. Thailand) was added. Vulcanizing additives i.e. ZnO, stearic acid, MgO, CBS, TMTD, ETU and Sulphur with content of 5, 2, 2, 1, 0.5, 0.5 and 2.2 phr respectively, was added consecutively. The obtained NR/CR compounds was then examined using Moving Die Rheometer (MDR, Gotech 2000, Thaiwan) for cure characteristic parameters.

2.2 Preparation of NR/CR foam: After masticating of NR/CR blends, it was incorporated with vulcanizing additives except for sulfur, 1 phr. of nano-silica and 10 phr. of blowing agent, which ADC or OBSH, was added respectively. After which sulfur was finally incorporated. The obtained compound were then processed in a compression molding (Chareon Tut Co. Ltd., Thailand) at 160 °C follow tc90 obtained from previous step.

2.3 Morphological Examination: Cell morphology of the foams were investigated using SEM (Hitachi, TM 3030, Japan) on gold sputtered razor cut foam sample under 15eV electron under 200 magnify investigation. Cell size and density were investigated from SEM micrographs.

2.4 Mechanical properties: Tensile test was performed using universal testing machine (Instron UTM, 5969, USA) follow ASTM D412. While compression set was performed follow ASTM D5016 by compressing specimen with the width 20 mm length of 20 mm and height of 15 mm to 50% of its original height. Then the compressed specimens were kept at 70 °C in an oven for 22 hours. After which, removed the specimen and left at room temperature for 30 minutes before measuring the height of the specimens. Then calculated percentage of compression set using the following equation:

$$\%compression\ set = \frac{H_0 - H_2}{H_0 - H_1} \times 100\%,$$

where H_0, H_2 and H_1 are initial, final and spacer height respectively.

2.5 Limiting oxygen Index and Flammability: LOI was carried out follow ASTM D2563-91 to evaluate the oxygen required in burning of test specimen. Burning rate was investigated follow Underwriters Laboratories 94V (UL-94). The test specimen was positioned as shown in Figure 1. Burning length is then noted. However, the experimental detailed and burn rating is explained elsewhere [5].

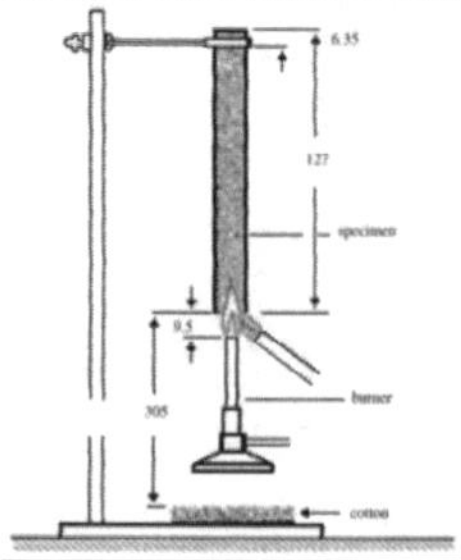

Figure 1. Schematic diagram shows the setting up of equipment for UL-94 testing [6].

IOP Conf. Series: Materials Science and Engineering 1280 (2023) 012020 doi:10.1088/1757-899X/1280/1/012020

3. Results and discussion

3.1 Morphology of NR/CR blend foam: cell morphology and cell size distribution are depicted in Figure 2-3. It was found that all foams contain close cell. For OBSH case, cell size was found to be smaller as CR content was increased. This could be due to polar interaction between nano-silica and CR could be performed between OH and Cl side group of nano-silica and CR resulting in well distribution of nano-silica. As silica acting as foam nucleating agent hence smaller foam cell size. In the case of ADC, cell size tends to be smaller than foam blown with OBSH. In fact the decomposition temperature of ADC is in the range of 180-210 °C and producing gas around 231 cm^3/g. of ADC at the highest temperature [7] whereas OBSH is decomposed at 140-160 °C producing gas around 120-150 cm^3/g of OBSH [7]. This could cause foam produced using ADC appeared to show smaller cell size. As CR content increased cell size of the foam increased nonetheless cell size was still smaller than foam produced using OBSH as blowing agent.

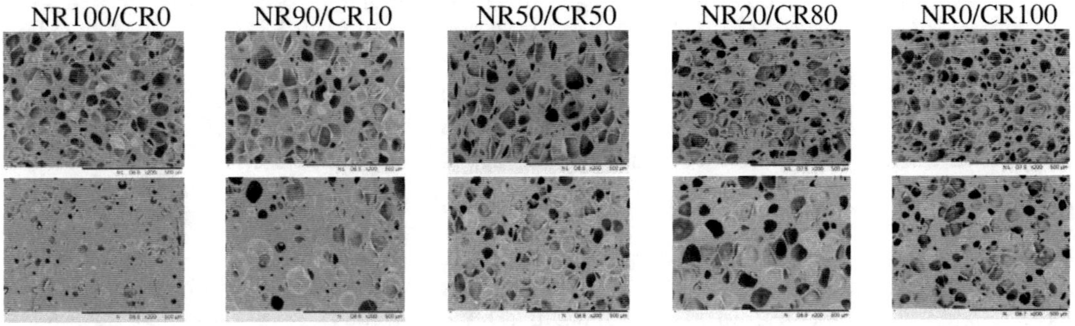

Figure 2. SEM micrographs of NR/CR blend foam blown with OBSH (upper row) and ADC (lower row) with CR contents.

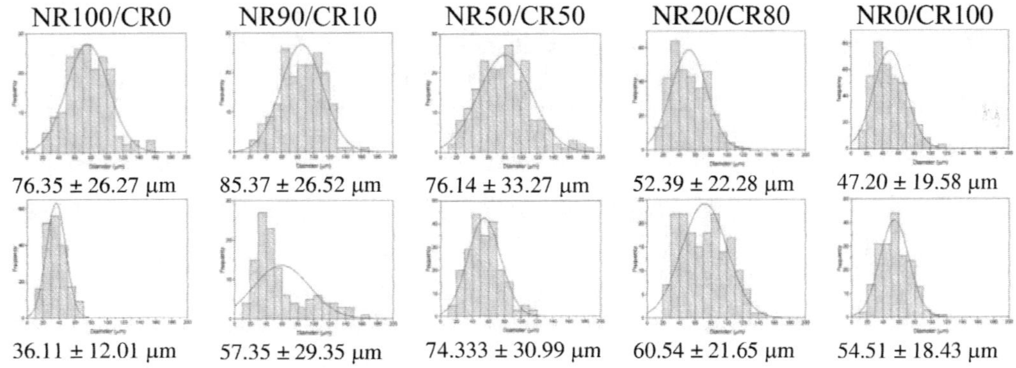

Figure 3. Cell size distribution of NR/CR blend foam blown with OBSH (upper row) and ADC (lower row) with CR contents.

3.2 Cell density and density of NR/CR foam: From SEM micrographs cell density and density of the foam could be evaluated and the results shown in Figure 4. The results in Figure 4 (a) show that cell density in the foam was increased with the content of CR. It was explained in our previous work [5] that the addition of silica could prolong cure time and the nano-silica could also well disperse in CR and consequently blowing agent resulting in decreasing cell size and increasing cell density. It can also be notice that when CR content is as high as 80 % cell density is increased significantly owing to the reason afore mentioned.

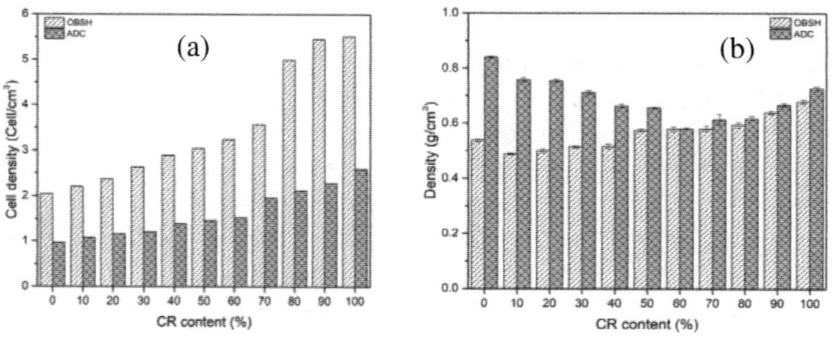

Figure 4. Cell density (a) and foam density (b) of NR/CR blend foam with various CR content and blown with OBSH compared with ADC.

In term of foam density, in Figure 4 (b), considering foam produced using OBSH, the density was increased with CR contents due to the density of CR is higher than NR. Whereas foam produced using ADC, foam with high NR content, from 100 to 50 %, exhibits high density. As shown in cell morphology and cell size distribution as cell size are small and cell density is low resulting in high density foam. When cell size and cell density was increased foam density was then decreased. It was not the case for blend with high CR contents, 60-100 % CR, in which density of CR affect density of the foam.

3.3 Tensile strength and Compression set (CS) of NR/CR blend foam:
Tensile strength and compression set of the foam are reported in Figure 5 (a) and (b). It was shown that tensile strength, in Figure 5 (a), of the foam is affected by foam morphology. As seen in SEM micrographs that foam blown with OBSH gave rise to larger cell size and cell density than ADC consequently weaken tensile strength of the foam. For compression set, CS, in Figure 5 (b), this property implies elastic recovery of materials. The lower the CS is the better elastic recovery. It can be seen from the results that at high NR contents CS tended to show permanent deformation. It should be concerned that during CS testing, the test specimens were kept at 70°C for 22 hours. This could cause thermal ageing especially in NR phase and this responsible for high compression set of the foam. Afore mention that CR possess higher thermal ageing resistance than NR thus foam with increased CR content exhibits lower CS. Considering foam with different blowing agents, foam produced from OBSH gave better elastic recovery. This could be explained by the higher cell density and acting as thermal insulator hence less thermal ageing during CS testing than another foam system.

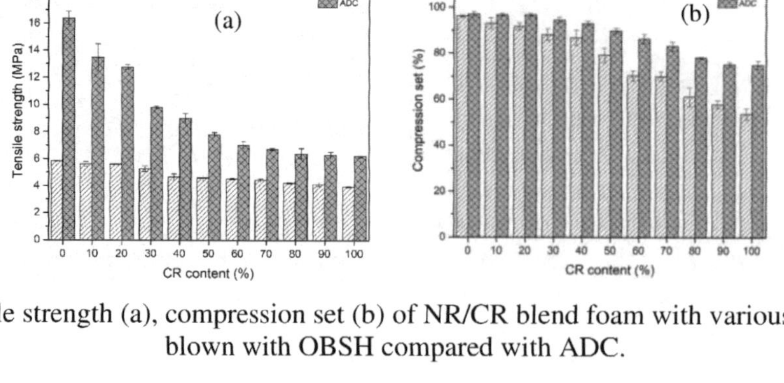

Figure 5. Tensile strength (a), compression set (b) of NR/CR blend foam with various CR content and blown with OBSH compared with ADC.

3.4 Flammability and UL-94 classification: Flammability is investigated in term of Limited Oxygen Index which is defined as the minimum concentration of oxygen in a mixture of oxygen and nitrogen that is needed to support the flaming combustion of a material and expressed in percentage volume. The results are shown in Figure 6. It could be seen that as CR content is increased % LOI was increased this is due to self-extinguish characteristic of CR. Cl attached to polymer main chain in every repeating unit could be released as HCl and cover polymer surface and protect polymer surface from oxygen. After dehydrochlorination, the carbonaceous matter was resulted to cover and protect polymer surface. The addition of CR into NR could slow down dripping of rubber which is due to the carbonaceous compound appeared after dehydrochlorination even though UL-94 rating was improved with CR content as high as 90 % in the blend.

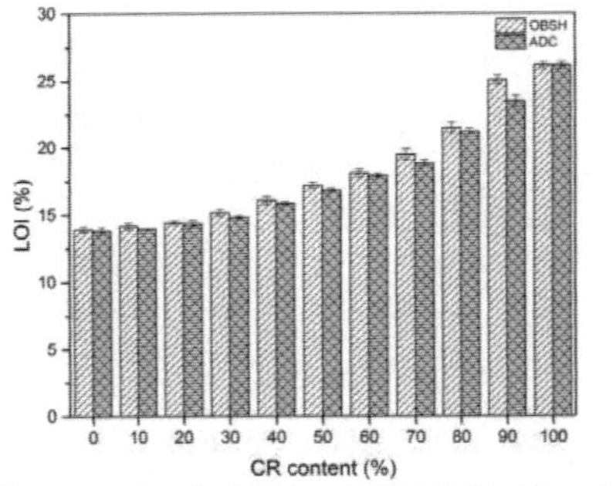

Figure 6. % LOI of NR/CR blend foam foaming with different blowing agent.

Table 1 UL-94 rating for NR/CR foam

NR/CR	UL-94 Classification	
Blowing agent	OBSH	ADC
100/0	V-2	V-2
90/10	V-2	V-2
70/30	V-2	V-2
50/50	V-2	V-2
30/70	V-2	V-2
20/80	V-1	V-1
10/90	V-1	V-1
100/0	V-0	V-0

4. Conclusions

OBSH and ADC are generally used as synergist blowing agents. In order to investigate the primary effects of these blowing agents on NR/CR blends they are used separately and compared. It was showed that using OBSH and ADC gave rise to foam with different cell morphology, cell size, cell size distribution, cell density and foam density. According to small cell size and cell density, foam produced using ADC could show better tensile strength whereas OBSH foam system could exhibit better thermal ageing and elastic recovery under compression set testing under prolong heating at 70 °C. Due to self-extinguish characteristic, NR/CR blend foam exhibit higher %LOI but foam with high density of air bubble could retard flame and hence need more oxygen to support flaming. For burning rate, it was found that incorporation of CR into NR could slow down dripping of rubber under flame although rating is slightly improved. It is expected that this type of foam can be applied for oil and heat aging resistance cellular rubber.

Acknowledgement

The authors are in debts to the Department of Materials science and Engineering, Faculty of Engineering and Industrial Technology, Silpakorn University for all support in the research and other related works.

References

[1] Lee E -K and Choi S -Y 2007 *Korean J. Chem. Eng.*. **24** 1070.
[2] Rostami-Tapeh-Esmaeil E, Vahidifar A, Esmizadeh E and Rodrigue D 2021 *Polymers* **13** 1565.
[3] South J T, Case S W and Reifsnider K L 2003 *Rubber Chem. Technol.* **76** 785.
[4] Ramesan MT, Alex R and Khanh N V 2005 *React. Funct. Polym.* **62** 41.
[5] Angsupanich, L 2022. *Master of Engineering Dissertation.* Department of Materials Science and Engineering, Graduate School, Faculty of Engineering, Silpakorn University, Sanamchandra Palace Campus, Nakornpathom Thailand.
[6] Wan L, Deng C, Zhao Z Y, Chen H and Wang Y Z 2020 *Strategy and Recent Progress Polymer* **12** 1-24.
[7] Rostami-Tapeh-Esmaeil E, Vahidifar A, Esmizadeh E and Rodrigue D 2021 *Polymer* **13** 1565.

IOP Conf. Series: Materials Science and Engineering 1280 (2023) 012021 doi:10.1088/1757-899X/1280/1/012021

Development of natural rubber foam by in-house Talalay method

H Klommueang[1] and W Smitthipong[1,2,*]

[1]Specialized Center of Rubber and Polymer Materials in Agriculture and Industry (RPM), Department of Materials Science, Faculty of Science, Kasetsart University, Bangkok,Thailand.
[2]Office of Research Integration on Target-Based Natural Rubber, National Research Council of Thailand (NRCT), Bangkok, Thailand.

*Corresponding author: fsciwssm@ku.ac.th

Abstract. The benefits of natural rubber foams (NRFs), include its buoyancy, sound absorption, supportability, and low weight. Generally, there are two main techniques to produce the NRFs: Dunlop method and Talalay method. The Dunlop technique is a cheap and conventional method, while the Talalay technique is a complicated method (cool down, vacuum and heating process). The objective of this research is to study the difference properties of NRFs between Dunlop and Talalay methods. Both techniques proposed the NRFs with similar foam density values. The mechanical properties of the NRF from in-house Talalay method were lower than those obtained by NRF from Dunlop method, may due to the in-house Talalay method generating the large bubble from the vacuum system. However, the FTIR spectrum of NRFs obtained from both methods showed the same functional groups.

1. Introduction

Natural rubber foams (NRFs) are foam products made from concentrated natural latex, they possess unique characteristics that make them possible for producing rubber goods like rubber foam [1-2]. Shoes, beds, pillows, and vehicle seats can be produced by rubber foam. Dunlop, one of the top business in the world for the production of rubber foam goods, was a company that created rubber foam for the first time [3]. The Talalay process is an alternative method for creating rubber foam, and it can result in softer rubber foams with greater flexibility than those made using the Dunlop procedure [4]. E.A. Murphy developed the Dunlop process, a key NRF production technique, in the late 1920s to produce latex from the sap of rubber plants. In this procedure, vulcanizing chemicals were combined with natural latex and whipped into froth with the help of a foam substance. The fluid foam was mixed with a delayed-action gelling agent, like sodium silicofluoride (SSF), and then pumped into a mould. The fluid foam was baked in a vulcanization oven after being allowed to gel into solid foam at room temperature. After baking, the foam was taken out of the mould, cleaned, and then heated a second time to eliminate any remaining moisture [5]. However, the Talalay method differs significantly from the Dunlop approach in several ways. The objective of this study is to compare the main properties of NRFs obtained from both conventional Dunlop and in-house Talalay methods.

2. Experimental

2.1 Materials

Concentrated natural latex (60% dry rubber content) from Num Rubber & Latex Co., Ltd., Trang, Thailand. Chemical agents consist of 10% potassium oleate (PO) solution, 50% sulphur (S) dispersion, 50% zinc diethyldithiocarbamate (ZDEC) dispersion, 50% zinc–2–mercaptobenzothiazole (ZMBT) dispersion, 50% Wingstay L (Wing L) dispersion, 50% zinc oxide (ZnO) dispersion, 33% diphenylguanidine (DPG) dispersion, and 12.5% sodium silicofluoride (SSF) dispersion. All the chemical agents from Thanodom Technology Co.,Ltd., Thailand.

2.2 Preparation of natural rubber foams (NRFs)

2.2.1 Dunlop method

The NRFs were prepared by removing any ammonia using a blending machine at 80 rpm for 1 min. The surfactant (PO) was then mixed, and the speed was increased to 160 rpm for 10 min or until the required volume was reached. The vulcanized group of (S, ZMBT, ZDEC) and Wing L were added and then blended for 1 min. Then, ZnO and DPG were mixed and blended for 1 min. Then, SSF was add and mixed for 3 min. Next the latex compound was poured into a mould. The sample should set for 45 min at room temperature. The foam was vulcanized at 90°C for 1.45 h. Finally, the NLF was then removed from the mould, washed to remove any residual chemicals, and dried at 70°C for 4 h.

2.2.2 In-house Talalay method

The NRFs were prepared by removing any ammonia using a blending machine at 80 rpm for 1 min. The surfactant (PO) was then mixed, and the speed was increased to 160 rpm for 10 min or until the required volume was reached. The vulcanized group of (S, ZMBT, ZDEC) and Wing L were added and then blended for 1 min. Then, ZnO and DPG were mixed and blended for 1 min. Next, the latex compound was poured into a mould. The mould was then sealed and placed in the temperature chamber, and the vacuum was decreased to 700 mmHg for 5 min to allow the foam to expand until the maximum capacity of the mould was reached. The temperature was then decreased to −30°C using a mechanical single-stage refrigeration system for 10 min to freeze the foam in the moulds. CO_2 gas was then fed into the system at a rate of 10 L/min for 5 min to initiate the gelling of the rubber foam. The foam was vulcanized by heating it to 90°C for 1.45 h. Finally, the NLF was then removed from the mould, washed to remove any residual chemicals, and dried at 70°C for 4 h.

Table 1. The raw materials composition of the natural rubber foams (NRFs).

Chemicals \ Sample	Wet weight (g)	
	Dunlop method	**In-house Talalay method**
NRL	166.67	166.67
PO	17.00	17.00
S	4.00	4.00
ZMBT	2.00	2.00
ZDEC	2.00	2.00
Wing L	2.00	2.00
ZnO	10.00	10.00
DPG	2.00	2.00
SSF	8.00	-

2.3 Characterizations

The NRFs' sample weight and measured volume were used to compute the foam density, which was estimated as follows:

$$D = \frac{M}{V} \quad (1)$$

where M is the weight of the rubber foam sample (kg) and V is the volume of the rubber foam sample (m^3).

Attenuated total reflection-Fourier transform infrared spectroscopy (ATR-FTIR, VERTEX 70) was used to examine the chemical functional groups of the foam samples between 500–4000 cm^{-1}

Compression stress of the NRFs was determined by a texture analyzer [2]. The NRF sample was 45 mm width × 45 mm length× 21.5 mm thickness, compressed at 75% from the foam surface at room temperature.

3. Results and Discussion

The foam density of the NRF samples is presented in Figure 1. We found that foam density of two types of NRFs are almost similar even Dunlop method and in-house Talalay method are quite different. The Dunlop method proposed the mixing technique to obtain the suitable bubble level, while the in-house Talalay method proposed different steps: mix, cool down, vacuum and heating process. NRF obtained from the Dunlop method exhibited heterogenous cell size [2,6] due to the mixing process, but NRF obtained from the in-house Talalay method may also represent even more heterogenous cell size based on the complicated method.

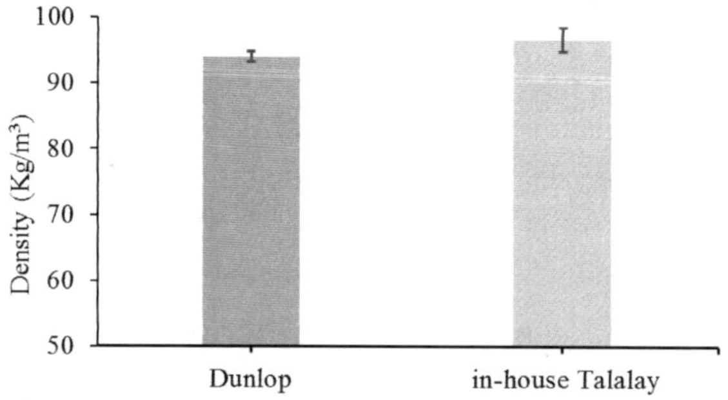

Figure 1. The density of the NRF samples.

Figure 2 shows mechanical properties of NRFs obtained from Dunlop and in-house Talalay methods. Both types of NRFs represented similar compression stress at plateau region (below 50% strain). However, the NRFs by Dunlop process had a higher compressive strength than the NRFs by in-house Talalay process at the solidification zone [6]. This may be due to the complex process of in-house Talalay method (cool down, vacuum, heat up) generating heterogeneous air bubble and thus it may result in reduction of its mechanical properties.

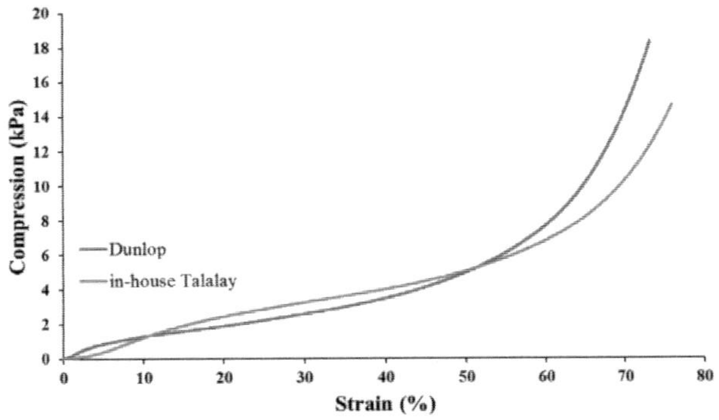

Figure 2. The Compression stress of the NRF.

The chemical compositions of both NRFs were studied by ATR–FTIR (Figure 3). There is no significant difference in the functional groups of NRFs. The functional group of natural rubber was found the of C-H stretching vibration at 2800-3000 cm^{-1} and the characteristic vibration of natural rubber representing C=C stretching at 1664 cm^{-1} [6]. Furthermore, this result also found that the bending vibration of C-H at 1300-1400 cm^{-1}.

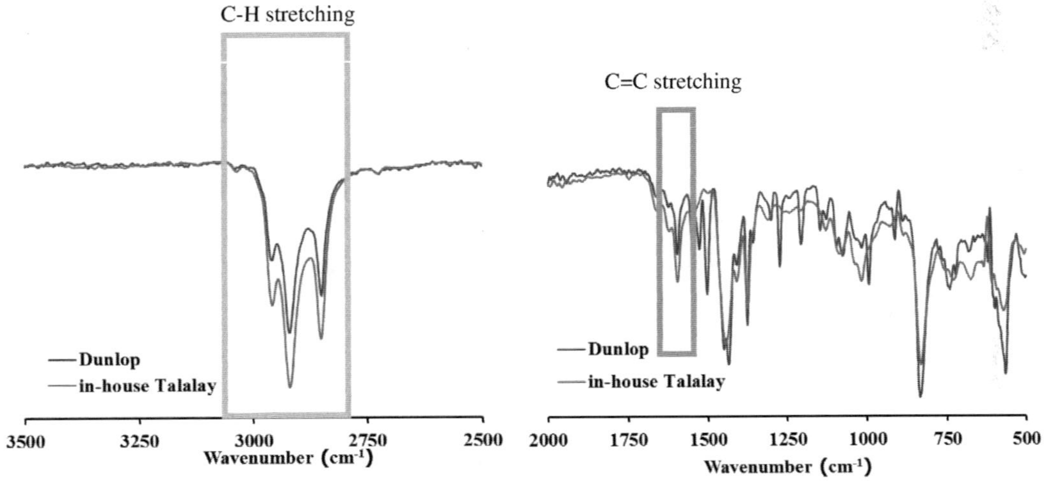

Figure 3. ATR–FTIR spectra of the NRF samples.

4. Conclusions

In this work, the natural rubber foams (NRFs) were prepared by two main techniques to produce: Dunlop method and in-house Talalay method. The foam density of the NRFs produced by the two methods was comparable. Due to the intricate testing procedure, the mechanical qualities of the in-house Talalay method were inferior to those achieved by the Dunlop approach. The FTIR spectrum revealed the same functional groups in the NRFs generated from both techniques.

Acknowledgement

Specialized Center of Rubber and Polymer Materials in Agriculture and Industry (RPM), Department of Materials Science, Faculty of Science, Kasetsart University, Bangkok, 10900, Thailand.

References

[1] Sukkaneewat B and Songkot S 2022 *Ultrason. Sonochem.* **82** 105873.
[2] Prasopdee T and Smitthipong W 2020 *Polym.* **12** 2745.

[3] Blackley D C 1997 *Polymer Latices* 229.
[4] Sirikulchaikij S, Nooklay B, Kokoo R and Khangkhamano M 2019 *Mater. Sci. Forum.* **962** 96.
[5] Phomrak S, Nimpaiboon A, Zhang B and Phisalaphong M 2020 *Polym.* **12** 1959.
[6] Suethao S, Ponloa W, Phongphanphanee S, Wong-Ekkabut J and Smitthipong W 2021 *Sci. Rep.* **11** 6097.

Incorporation of Rubber Waste to Fiber Cement Composite: Comparative Study of Rubber Tire Waste and Rubber Band Waste

S Kajon[1], K Hancharoen [2], P Kamhangrittirong[2] and P Suwanna[1,*]

[1] Department of Materials Science, Faculty of Science, Kasetsart University, Bangkok 10900, Thailand.
[2] Center of Building Innovation and Technology (CBIT), Department of Building Innovation, Faculty of Architecture, Kasetsart University, Bangkok 10900, Thailand.

*Corresponding author: pimsiree.s@ku.ac.th

Abstract. This study compares the effects of rubber tire waste and rubber band waste on physical, mechanical, and thermal properties of cement composite for roofing application. The rubber particles, prepared from waste rubber tire (R_T) and/or calcium carbonate filled rubber band (R_B), were mixed with cement paste and tested for workability and flexural strength. It was found that R_T yields higher workability than R_B when mixed with the cement paste. The rubber-cement composites (CR) with the combination of R_T and R_B at the R_T:R_B volume fraction of 1.00:0.00 (CR_T), 0.75:0.25 ($CR_{T.75/B.25}$) and 0.50:0.50 ($CR_{T.50/B.50}$) have sufficient workability and flexural strength. When mixing different rubber particles with oil palm fibers and the cement paste to obtain CFR_T, $CFR_{T.75/B.25}$, $CFR_{T.50/B.50}$ rubber-fiber-cement composites, the composites with higher R_T content have higher flexural strength, while those with higher R_B content have lower thermal conductivity, implying better thermal insulation property. More importantly, all three types of rubber-fiber-cement composites yield excellent thermal insulation property, and meet the flexural strength required by the industrial standard for the fiber-cement roof tile sheets. The results suggest that the waste rubber-waste oil palm fiber-cement composites can be viable candidates for roofing application with promising potential for energy saving.

1. Introduction

A roof is an important part of a building because it protects the building from rain, wind, and heat from solar radiation. If the building's roof has sufficient thermal insulation property, it can save energy for cooling within the building by decreasing the heat passing through and/or collecting in the layer under the roof. Many reports have demonstrated improvement of thermal and/or sound insulation properties of cementitious composites with either natural fiber or waste tire rubber [1-6]. Our previous studies have also shown that adding waste natural fibers in combination with waste rubber tire particles to the cement matrix can produce greater advantages than adding each constituent separately [7-8]. These advantages include the enhancement of flexural strength as well as thermal and sound insulation properties. Even though the water insulation property of these rubber-fiber-cement composites was slighlty lower than that of the rubber-cement composites, it was still much better than that of the fiber-cement composite. Therefore, the rubber-fiber-cement composite is of great interest as a potential material for an energy-efficient roofing system.

In this work, we investigate the properties of the waste oil palm fiber composites with two different kinds of rubber particles, namely the waste rubber tire and waste rubber band particles. The rubber waste effects on workability, flexural strength, and thermal conductivity of the cement composite for roofing application are analyzed and compared. To be more definite, workability here indicates the ease and homogeneity the freshly mixed cement paste composite that can be mixed, placed, consolidated, and finished.

2. Experimental

2.1 Materials

Waste rubber tire particles from truck scrap tires and waste rubber band particles with the size of 30 mesh (0.420 - 0.595 mm) were purchased from Q2 Thailand Co., Ltd, Thailand and Kungvarl Industrial Rubber Co., Ltd, Thailand, respectively. In their manufacturing processes, the rubber tire had been filled with carbon black, while the rubber band with calcium carbonate. Oil palm fibers used in this research were waste oil palm empty fruit bunches obtained from Suksomboon Palm Oil Co., Ltd., Thailand. High early strength portland cement (PC) type I was purchased from TPI PLC Co., Ltd. Bacterial nanocellulose (BNC) in the form of coconut jelly containing bacteria strain *Gluconacetobacter xylinus*, purchased from Theppadungporn Coconut Co., Ltd., Thailand, was reproduced and used to coat the oil-palm fibers. Other chemicals used to prepare the rubber tire particles before adding to the cement mixture were supplied by Chemipan Corporation Co., Ltd. Thailand; these included sodium hydroxide (NaOH, industrial grade, Taiwan), potassium permanganate ($KMnO_4$, industrial grade, China), sulfuric acid (H_2SO_4, industrial grade, China) and sodium laureth sulfate (SLES, industrial grade, Germany).

2.2 Preparation of waste rubber particles

The surface of waste rubber tire is typically non-polar and hydrophobic; hence, the waste rubber tire particles do not mix well with water. It is necessary to treat the surface of the waste rubber tire particles to achieve a homogeneous mixture with the cement paste [9]. First, the rubber tire particles were cleaned by soaking in a 1% NaOH solution for 24 h and washed with clean water. Then, they were oxidized with a 5% $KMnO_4$ solution for 1 h in a water bath at 60°C, while the pH value was maintained at 2-3 using a 6 mol solution of H_2SO_4 acid.

The rubber particles were then rinsed with clean water, soaked in a 1% sodium laureth sulfate solution at 60°C for 1 h to complete the sulphonation reaction, and finally dried at 60°C for 24 h. For the waste rubber band, the particles were used as received because they mixed relatively well with the cement paste without any treatment.

2.3 Preparation of waste oil-palm fibers

The oil palm fibers were processed from the empty fruit bunches by chopping and sieving through a 30-mesh sieve. The average fiber length was approximately 10 mm. To reduce water absorption during mixing with the cement paste, the fibers were then coated by BNC [10]. First, the BNC in the form of coconut jelly sheet was reproduced by culturing the bacteria strain *Gluconacetobacter xylinus* in a mixture of sugar and coconut juice in the ratio of 1:100 by weight with small amount of vinegar (5% acetic acid) to maintain pH ~ 4 for seven days at room temperature. To prevent overaging, the growth of BNC was impeded by washing it with 1% w/v NaOH solution at 80°C for 1 h, then with clean water until the pH value of 7-8 was obtained. The resultant BNC was kept in a refrigerator (4°C) until further use. To coat the oil palm fibers, the BNC was first cut into pieces of approximated dimensions 1 cm × 1 cm and mixed with water in the 1:100 weight ratio, and then the fibers were added into the mixture. The container was covered with a thin cloth for 24 h to allow the BNC to coat on the fibers. Finally, the coated oil palm fibers were filtered out, washed with tap water, and dried in a hot air oven at 60°C for 24 h and stored in a desiccator.

2.4 Specimen preparation

In this study, there were eight specimens with different compositions as summarized in Table 1. The main ingredients were high early strength portland cement type I (C), and water (W) (with W/C 0.5), and the additional ingredients were the modified waste rubber tire particles (R_T) and/or the waste calcium carbonate filled rubber band particles (R_B) and in some cases, the BNC coated oil-palm fibers (F). Note that the numbers after letters T and B indicate the volume fraction of R_T and R_B, respectively, in each formulation. For example, $R_{T.75/B.25}$ means the volume fraction R_T:R_B is 0.75:0.25. However, because the density of R_B was approximately 0.67 times lower than that of R_T, the weight of R_T and R_B in each formulation was recalculated accordingly to obtain the proper volume fraction of the rubber particles. Mixing procedures were performed for all specimens in a rotary mixer. First, water was added to the cement powder and mixed at a low speed for 2 min. Then, rubber particles were added and mixed for 3 min. When applicable, the coated fibers were added and mixed for another 5 min. Specimens were casted into their molds (rectangular box of dimensions 200 mm × 200 mm × 10 mm) and demolded after 24 h. Afterward, specimens were cured in a water bath at room temperature for 28 days.

2.5 Materials characterization

The workability of the fresh cement paste composites was measured using a flow table method according to ASTM C1437. The specimen flexural strength was determined according to ASTM C293 (center-point loading) using a universal testing machine (Chun Yen Testing Machines, model CY-6040A4, Taiwan) and was expressed as modulus of rupture (MOR). The specimen thermal conductivity was measured according to ASTM C518 using a thermal conductivity tester (EKO Instruments, model HC074, Japan).

Table 1. Mixing compositions for the cement composites.

Formula	High early strength portland cement, C [g]	Oil-palm fiber, F [g]	Waste tire rubber, R_T [g]	Waste calcium carbonate filled rubber band, R_B [g]	W/C ratio
CR_T	100.0	-	50.0	-	0.5
$CR_{T.75/B.25}$	100.0	-	37.5	8.3	0.5
$CR_{T.50/B.50}$	100.0	-	25.0	16.7	0.5
$CR_{T.25/B.75}$	100.0	-	12.5	25.0	0.5
CR_B	100.0	-	-	33.0	0.5
CFR_T	100.0	10.0	50.0	-	0.5
$CFR_{T.75/B.25}$	100.0	10.0	37.5	8.3	0.5
$CFR_{T.50/B.50}$	100.0	10.0	25.0	16.7	0.5

3. Results and discussion

The first part of this work was to compare the workability and flexural strength among the rubber cement composites and select formulations that can be further developed into rubber-fiber-cement composite roof tiles. According to our preliminary results, the fresh rubber-cement paste composite should have workability of at least 18-20% to allow addition of oil palm fibers. Furthermore, when the composite is fully cured, it should have flexural strength of at least 18 MPa to satisfy the industrial standard for the fiber-cement roof tile sheets (TIS. 1427-2540).

Figure 1 shows the workability of the fresh rubber-cement composites. The results show that the CR_T composite had the highest workability at 29.96 ± 0.19%, making it easiest to mix and pour into the mold. The workability of the composite decreased with the R_B content and was minimum in the CR_B composite at the value of 13.21 ± 0.39%. R_T yields better workability because its surface was treated as described in the experimental section to achieve a homogeneous mixture with the cement paste. Without such treatment, R_T did not mix well with cement paste as shown in our previous study [7-8]. In contrast, the surface of R_B was not treated, so it yields lower workability than the treated R_T. However it still mixes with the cement paste better than the untreated R_T. The reason for this needs further investigation, but it is hypothesized that calcium carbonate in the rubber band plays an important role. If the untreated R_B can be used to replace R_T even partially, the chemical waste from the rubber treatment process will be reduced.

The flexural strength of all rubber cement composites is presented in Figure 2, reporting the flexural strengths of CR_T, $CR_{T.75/B.25}$, $CR_{T.50/B.50}$, $CR_{T.25/B.75}$, and CR_B composites at 18.91 ± 0.52, 22.55 ± 2.06, 19.91 ± 0.48, 16.89 ± 0.85 and 18.41 ± 0.38 MPa, respectively. While the effect R_T:R_B volume fraction on the flexural strength does not exhibit a clear trend, but it provides information for screening of the composite formulations for second part of this study. When considering both the workability and the flexural strength, the rubber-cement composites that satisfy both criteria were CR_T, $CR_{T.25/B.75}$ and $CR_{T.50/B.50}$. Hence they were selected for further development as the rubber-fiber-cement composite roof tiles.

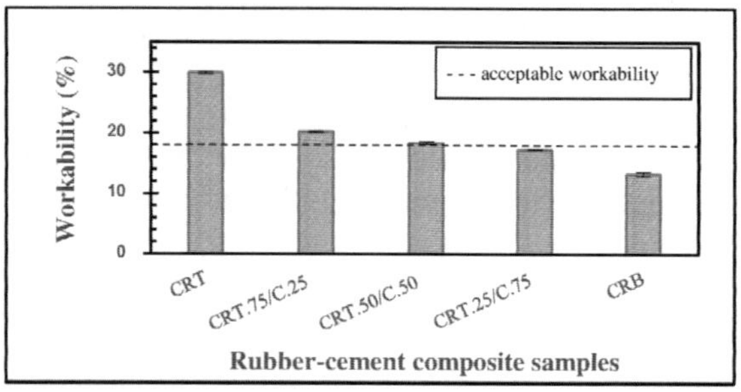

Figure 1. Workability of fresh rubber cement composites

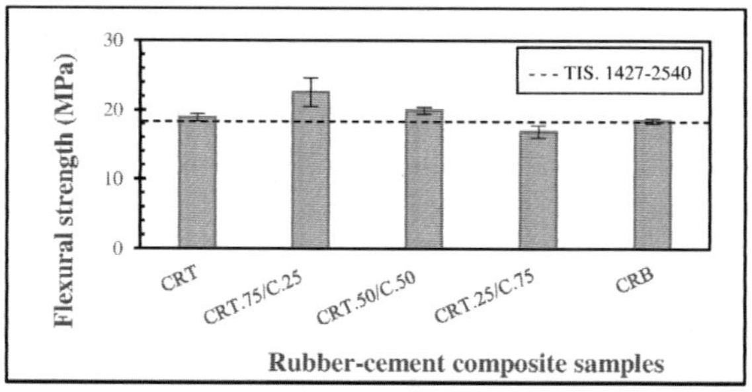

Figure 2. Flexural strength of fully cured rubber cement composites (28 days)

For roofing application, the composites with higher flexural strength and/or lower thermal conductivity (and hence better thermal insulation property) are more desirable. As shown in Figure 3, the flexural strengths of the CFR_T, $CFR_{T0.75/B.25}$ and $CFR_{T.50/B.50}$, rubber fiber cement composites were 24.81 ± 0.28, 22.19 ± 0.44 and 19.54 ± 0.54 MPa, respectively, which were approximately 37.83, 23.28 and 8.56 % higher than the industrial standard. The flexural strength was highest in CFR_T and was found to decrease with the R_B content. This is probably because the treated R_T particles mixed better with the cement paste than the untreated R_B, resulting in better interface between the rubber particles and the cement, and consequently higher flexural strength in CFR_T.

The thermal conductivity of the composite specimens is reported in Figure 4 in comparison with those of the commercial fiber-cement roof tiles made in Thailand, which have thermal conductivity in the range of 0.39 - 0.41 W/m·K. The thermal conductivities of CFR_T, $CFR_{T.75/B.25}$ and $CFR_{T.50/B.50}$ were 0.3894 ± 0.0851, 0.2853 ± 0.0557 and 0.2180 ± 0.0457 W/m.K, respectively. The thermal conductivity of the composite decreases with the R_B content, because R_B has lower thermal conductivity than R_T. While the CFR_T composite had comparable thermal conductivity to the commercial tiles, the $CFR_{T.75/B.25}$ and $CFR_{T.50/B.50}$ had thermal conductivities of at least 26.85 and 44.10 % lower. This reduction of the thermal conductivity, which can be considered as an improvement of thermal insulation property, is highly desirable and can be achieved by partially replacing the waste rubber tire particles with the waste rubber band particles.

When comparing the effects of the two types of rubber particles on the properties of the composites, the waste rubber tire particles were found to give better workability and flexural strength while the waste rubber band particles were found to yield better thermal insulation property.

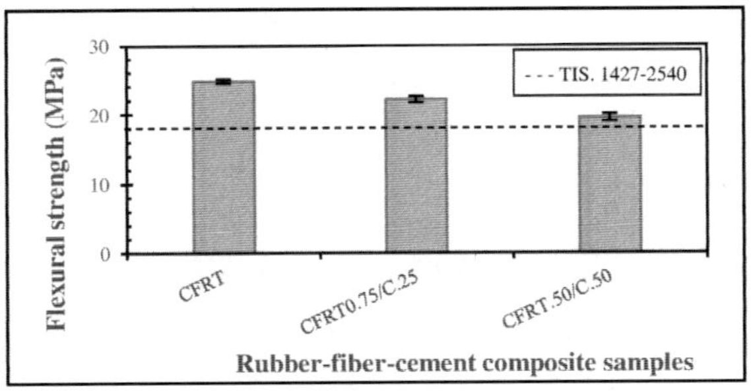

Figure 3. Flexural strength of rubber fiber cement composites

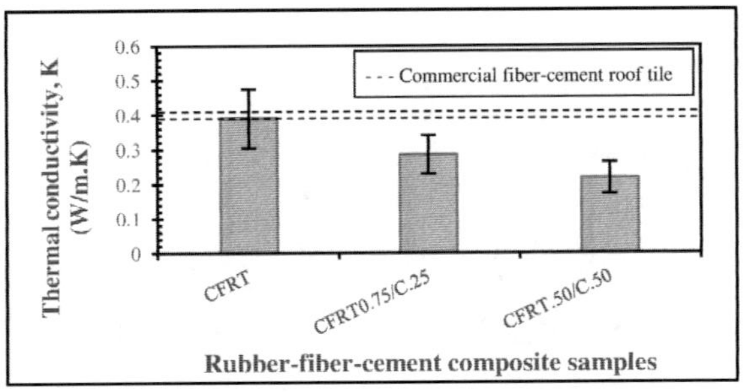

Figure 4. Thermal conductivity of rubber fiber cement composites

4. Conclusions

This research compared the effects of waste rubber tire particles (R_T) and waste calcium carbonate filled rubber band particles (R_B) on workability, flexural strength, and thermal conductivity of the cement composites. It was found that when mixed with the cement paste, R_T yields better workability than R_B. In the rubber-oil palm fiber-cement composites (CFR_T, $CFR_{T0.75/B0.25}$, $CFR_{T0.50/B0.50}$), those with higher R_T content have higher flexural strength, while those with higher R_B content have lower thermal conductivity (better thermal insulation). All three types of rubber-fiber-cement composites yield excellent thermal insulation property and satisfy the flexural strength required by the industrial standard for the fiber-cement roof tile sheets. The results suggest these composites can be viable candidates for roofing application with promising potential for energy saving. The $CFR_{T0.75/B0.25}$ composite is especially attractive as it has well-balanced combination of mechanical and thermal properties (flexural strength approximately 23.28% higher than the industrial standard and thermal conductivity of at least 26.85% lower than the commercial fiber-cement roof tiles). Further investigation should be conducted, especially at the level of real application.

Acknowledgement

This research is financially supported by the National Research Council of Thailand under the project "Innovation Management of Natural Rubber Products in Building Materials and Composite Materials for Sustainable Well-Being Development" (fiscal year 2020).

References

[1] Elwin S R, Abarna K, Aiswarya A, Anuruthra V and Atchaya A 2019 *Int. J. of Adv. Research in Basic Eng. Sci. Technol.* **5** 102.
[2] Hall M R, Najim K B and Hopfe C J. 2012 *Appl. Therm. Eng.* **33** 77.
[3] Khaloo AR, Dehestani M and Rahmatabadi P 2008 *J. Waste Manag.* **28** 2472.
[4] Michels C, Güths S, Marinoski D L and Lamberts R 2021 *Energy Build.* **231** 110569.
[5] Paine K A and Dhir R K 2010 *Const. Mater.* **163** 7.
[6] Piti S 2009 *Con. Build. Mat.* **23** 1084-.
[7] Hancharoen K, Kamhangrittirong P and Suwanna P 2020 *Key Eng. Mater.* **861** 465.
[8] Hancharoen K 2020 Processing innovation and properties of cement composite roofing tiles incorporating nanocellulose coated fiber and rubber tire waste, Ph.D. Dissertation, Kasetsart University.
[9] Mohammadkazemi F, Doosthoseini K, Ganjian E and Azin M 2015 *Constr. Build. Mater.* **101** 958.
[10] He L, Ma Y, Liu Q and Mu Y 2016 *Constr. Build. Mater.* **120** 403.

IOP Conf. Series: Materials Science and Engineering 1280 (2023) 012023 doi:10.1088/1757-899X/1280/1/012023

Artificial Neural Network Model for Predicting the Energy Loss of Natural Rubber Foam

P Kaewpradit[1*], N Uthaipan[2] and C Dechwayukul[3]

[1] Department of Chemical Engineering, Faculty of Engineering, Prince of Songkla University, Bangkok 90110, Thailand
[2] Rubber Technology and Engineering program, International College, Prince of Songkla University, Bangkok 90110, Thailand
[3] Department of Mechanical Engineering, Faculty of Engineering, Prince of Songkla University, Bangkok 90110, Thailand

*Corresponding author: pornsiri.k@psu.ac.th

Abstract. In this work, a back-propagation artificial neural network model was optimally developed based on 25 experimental datasets for predicting the energy loss percentage of natural rubber foam. The foam specimens were prepared in a Banbury internal mixer at various conditions of mixing temperature (40-80°C), rotor speed (40-80 rpm), and mastication time (1-5 min). Stress-strain loops were analyzed by applying compressive force at aspeed test of 500 mm/min with the capacity of load cell 2.5 kN, the energy loss was further calculated. In model development, the experimental datasets were randomly divided into 70:15:15 for training, validation, and testing respectively. Levenberg–Marquardt algorithm was used as a training function was used because of its fast convergence. The prediction results revealed that the average prediction accuracy of the three models is higher than 90%. From a material design point of view, the developed model could be implemented to find the proper mixing conditions to obtain the material with the maximum energy dissipation.

1. Introduction

The final properties of the rubber compounds can be varied depending on the operating conditions of the Natural Rubber (NR) blending process, such as mixing temperature, mixing speed, and mastication time in the internal mixer. Artificial Neural Network (ANN) model has been widely developed in the literature to predict rubber properties to ensure desired quality in rubber products [1-3]. In their works, the average prediction accuracy of the ANN model ranged from 85% to 97%, which was related to the quality of the experimental data in the model training as well as the complexity of the process.

Most research in the literature has focused on the development of models for predicting the properties of solid specimens. There are rare studies in the application to predict the lost energy of the closed-cell foam. Although it is generally known that foam has a better impact-absorbing property than solid samples [4-6]. In this study, Back-Propagation Artificial Neural Network (BP-ANN) model has been developed and improved to predict the percentage of the lost energy of NR foam. The prediction model may be useful in an advanced or smart material design perspective.

2. Experimental

Standard Thai Rubber (grade STR20) was initially masticated for 1, 3, and 5 min (t_{mast}) at mixing temperatures of 40, 60, and 80°C (T_{mix}) and a rotor speed of 40, 60, 80 rpm (R_{mix}) in the internal mixer (Thermo Scientific HAAKE™ PolyLab™ OS system, Germany). The chemical additives were then added sequentially, including 10 phr precipitated silica, 10 phr china clay, 5 phr paraffinic oil, 1.7 phr ZnO, 0.5 phr steric acid, 1.5 phr ionoal ®LC. and 5 phr DNPT. The filling factor was limited to 0.7 and the total mixer volume was 379 ml. Curing agent (1.5 phr sulphur) and accelerator (0.3 phr TMTD, and 1 phr TBBS) were added to the compound using a two-roll mill at room temperature. The obtained compound was molded with dimensions of width 15cm×length 15cm×thickness 1cm by a hydraulic press machine (SiamLab SLLP-30A) at 150°C for 12 min. In this study, 180 g (80% of mold volume) of the rubber compound was vulcanized.

For the hysteresis test, the sample was die-cut into a cylindrical disc with a diameter of 28.6 mm ± 0.1 mm and a thickness of12.5 ± 0.5 mm. The three stress-strain loops were obtained by applying a compressive load at a500 mm/min velocity test with a 2.5kN load cell capacity (Universal testing machine ZwickRoell Model Z005, Germany). The specimens were tested at 65% maximum strain. Loading energy and unloading energy were calculated by Eq. (1), lost energy was the difference between the loading energy and unloading energy.

$$W = \int_0^{\varepsilon_u} \sigma_i(\varepsilon) d\varepsilon \tag{1}$$

where $\sigma_i(\varepsilon)$ is the loading or unloading curvilinear functions, and ε_u is the total strain at the unloading point. The hardness of the samples was measured on the surface of the cured specimens (minimum thickness of 6 mm) with a Shore OO durometer (BAREISS Test Stand BS61-II, Germany) following the ASTM D2240-05. The compression and deformation tester, Gotech GT-7049 (Taiwan) was used to determine the compression set. The foaming specimen was compressed to 50±2% of its original thicknesses and was oven-baked at 70°C for 22 h (ASTM D 1056-07). The percentage of the compression set can be calculated as follows:

$$\text{Compression set, \%} = \left[\frac{t_0 - t_1}{t_0 - t_s}\right] \times 100 \tag{2}$$

where t_0 is the original thickness (12.5 mm), t_1 is the thickness of the specimen after the specified recovery period (mm) and t_s is the height of spacer bar used.

3. Results and discussion

3.1. Experimental results

The experimental results in Table 1 show the effect of the mixing temperature (T_{mix}), the rotary speed (R_{mix}), and the mastication time (t_{mast}) on the density, hardness, compression set, and %lost energy of the rubber foam. The results show that all mixing parameters have a significant effect on foam density, hardness, compression set, and lost energy. The properties have the density of 550-700 g/cm^3, the hardness of 70-80 Shore OO, the compression set of 40-50%, and the lost energy of 16-30% within the experimental range. Effect of the mixing parameters on %lost energy was shown in Fig.1. The results show that T_{mix} increase, %lost energy increases and decreases; R_{mix} increases, %lost energy increases but t_{mast} increases, %lost energy decreases.

Table 1. Effect of mixing parameters on foam properties (range values)

Properties	T_{mix} (°C) 60 rpm, 5 min	R_{mix} (rpm) 60°C, 5 min	t_{mast} (min) 60°C, 60 rpm
Density (g/cm^3)	662-705	549-628	652-677
Hardness (ShoreOO)	79-81	70-75	71-80
Compression set (%)	38-41	45-53	42-45
Lost energy (%)	16-21	22-29	26-29

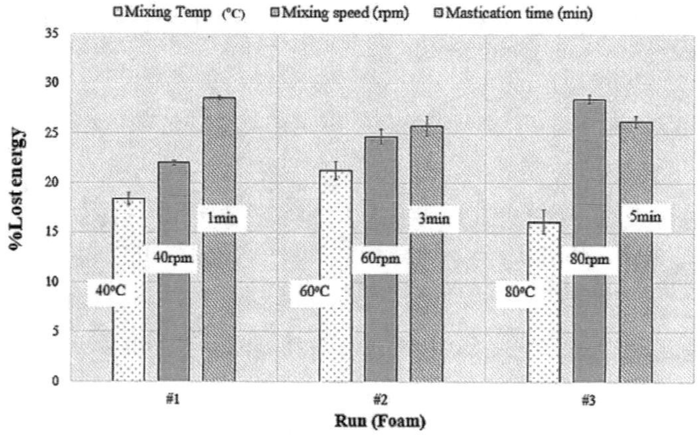

Figure 1. Effect of mixing parameters on %lost energy

3.2. Prediction results

Back-Propagation Artificial Neural Network (BP-ANN) model was created through MATLAB R2021b software using 25 experimental datasets. The datasets for training, validation, and testing were randomly divided into 70:15:15. Levenberg–Marquardt algorithm was used as a training function due to its rapid convergence [7]. In the structure of PB – ANN, one hidden layer was selected with a nonlinear Tan-Sigmoid transfer function (f_h = *tansig*), and a linear transfer function (f_o = *purelin*) was applied in the output layer (Fig.2).

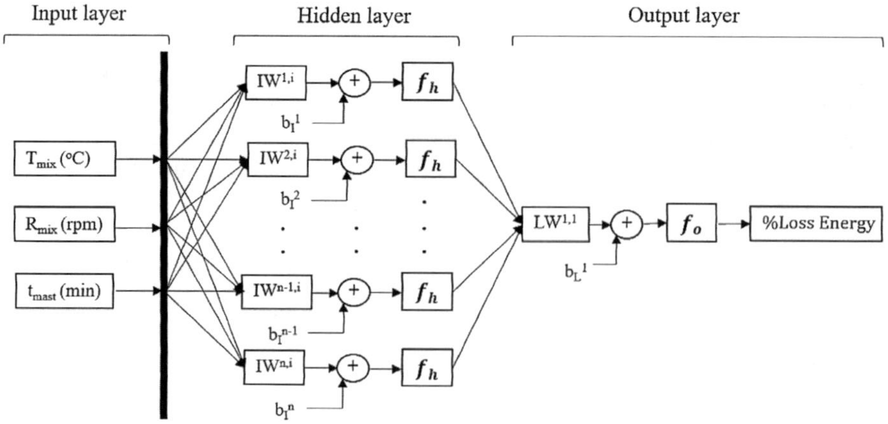

Figure 2. Structure of BP-ANN model.

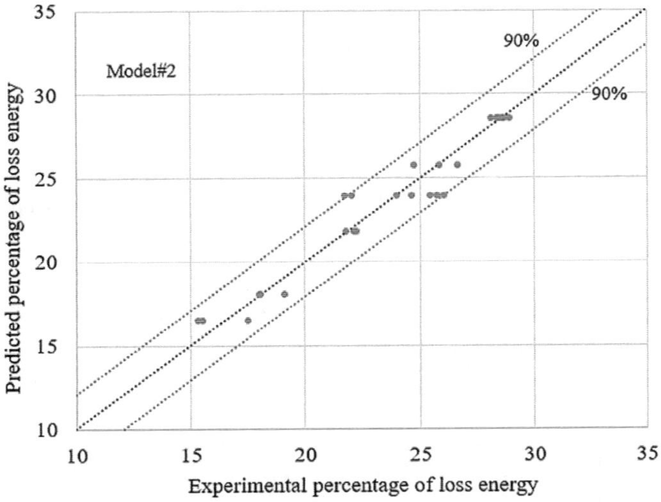

Figure 3. Comparison of predicted and experimental output.

Mean Square Error (MSE) was applied as a model performance function. The ANN model structure was chosen optimally by varying the number of hidden nodes from 2 to 20. In this case, the structure was chosen as 3-8-1 (input-hidden-output nodes) since of the lowest MSE. The average prediction accuracy was 96%.

3.3. Relative importance of parameters

The relative importance of mixing parameters was further analyzed using two different methods such as Olden's and Garson's algorithms [8-9]. These methods were used to evaluate the influence of multiple significant variables based on the weighting matrixes of input-hidden and output-hidden obtained from the BP-ANN model.

Table 2. Relative index by Olden and Garson methods

Methods	Olden	Garson
T_{mix} (°C)	26.8% (3)	27.2% (3)
R_{mix} (rpm)	34.3% (2)	33.8% (2)
t_{mast} (min)	39.0% (1)	39.0% (1)

Table 2 shows the evaluation results. It has been found that both algorithms give the same results which t_{mast} has the highest impact on %loss energy or energy dissipation. It has also been reported in [10] that the elasticity of rubber decreases or tan δ decreases when the mastication time increases.

4. Conclusions

In this research, Back-Propagation Artificial Neural Network model has been developed and improved to predict the percentage of the lost energy of natural rubber foam. The model has been optimally developed based on 25 experimental datasets. In this case, the model structure has been chosen as 3-8-1 (input-hidden-output nodes) with 96% average prediction accuracy. Relative importance of mixing parameters has been further analyzed using Olden's and Garson's algorithms to evaluate the influence of multiple significant variables. The evaluation results show that mastication time has the highest impact on %loss energy or energy dissipation of the foam specimen.

Acknowledgement
This work was funded by National Research Council of Thailand. The authors would like to thank Faculty of Engineering, and International college, Prince of Songkla University for all their support.

References

[1] Jin W, Liu Y and Gao Z 2017 *J. Appl. Polym. Sci.* **134** 45391.
[2] Kopal I, Labaj I, Harnicarova M, Valicek J and Hruby D 2018 *Polym.* **10** 644.
[3] Ghaffarian N and Hamedi M 2020 *IJE Transactions B: Applications.* **33** 2319.
[4] Fuss F K 2015 *Procedia Eng.* **112** 116.
[5] Fuss F K, Belbasis A, van den Hazel B and Ketabi A 2015 *Proc. Technol.* **20** 98.
[6] https://essay.utwente.nl/69908/1/CONF-Report_internship%20Bennie%20van%20den% 20 Hazel. pdf;March 2022.
[7] Xiang K L, Xiang P Y and Wu Y P 2014 *Mater. Des.* **57** 180.
[8] Sahu D, Chaurasia R C and Suresh N 2019 *Int. J. Coal Prep.* Util. **42:3** 389.
[9] Demirbay B, Kara D B and Ugur S 2022 *Expert Syst. Appl.* **207** 117937.
[10] Hayeemasae N, Waesateh K, Soontaranon S and Masa A 2020 *J. Elastomers Plast.* **53** 1.

The Study of Electrical Resistance of Silver Nanowires in Isopropanol for Heater Application

R Jaisutti[1], N Indrarit[1], K Pattanarat[1], T Khamla[2] and N Petchsang[2,3]*

[1] Department of Physics, Faculty of Science and Technology, Thammasat University, Khlong Nueng, Khlong Laung, Pathumthani 12121, Thailand.
[2] Department of Materials Science, Faculty of Science, Kasetsart University, 50 Ngam Wong Wan Road, Lat Yao Chatuchack Bangkok 10900, Thailand.
[3] Specialized Center of Rubber and Polymer Materials for Agriculture and Industry (RPM), Faculty of Science, Kasetsart University, Bangkok 10900, Thailand.

*nattasamon.p@ku.ac.th:

Abstract. Silver nanowire is one of the most attractive materials for making devices. In this work, we are interested in using silver nanowires for heater application. To get a good heater, the electrical resistance of the device should be low in order to generate a high temperature. The media to disperse nanowires is also important and helps inter-connection between nanowires. Hence, we study the different amounts of isopropanol (IPA) in water as the media to disperse silver nanowires. The nanowire solutions were prepared under different percentages of isopropanol varying from 0% to 100%. Optical images showed the dispersity of nanowires after the media evaporated. The electrical resistances of spun silver nanowires on glass slides were conducted. The results show that the electrical resistance of nanowires in different amounts of IPA was separated into 2 groups. First, from 0% to 50% of IPA, the electrical resistances were lower than 15 Ohms. Second, for 75% to 100% of IPA, the electrical resistances show more than 85 Ohms. After heater testing of the first group, the temperature is stable and repeatable with a certain applied voltage and current.

1. Introduction

Silver nanowires (Ag NWs) are a good candidate to create a device with high performance on heater applications [1-6]. It is highly conductive, and after applying a voltage, heat can be produced as in the equation below [7-8].

$$Q = \frac{V^2}{R_S} t$$

Where Q is heat (J), I is current (A), V is voltage (V), R is sheet resistance (W/sq), and t is time under applied voltage (s). Silver nanowires are suitable materials to fabricate the heater, and in order to make a good-performance heater, the electrical resistance should be as low as possible. Here in this work, we investigated the electrical resistance of the heaters under different concentrations of isopropanol (IPA) in water to find the optimum IPA concentration that enhances the heater's performance.

2. Methodology

Synthesized Ag nanowires: The Ag NWs were synthesized by using the polyol method. Briefly, Ethylene glycol (EG) 40 mL and PVP (MW=50,000) 0.22 g were put in a three-neck flask. After heating the temperature to 170 °C, AgCl 2 mg was added and waited for 20 min. At the same time, $AgNo_3$ 0.3 g was dissolved in EG in a separate vial. The solution was then added to the tree neck flask, and the silver nanowires appeared within 10 min. Once complete, the nanowires solution was washed and centrifuged with acetone and isopropanol two times before storing them in IPA.

Heater preparation: A 2%w/w of silver nanowires was prepared and used as a stock solution. A glass slide was cleaned with water and followed by acetone two times. Then it was further cleaned by a plasma cleaner for 20 s. After that PEDOT: PSS was spun as the first layer, followed by 600 µL of the nanowire stock solution two times. A gold electrode was applied on both edges of the glass slide. The current and voltage were applied using a Keithley electrometer, and temperatures were measured with a DS1820 sensor.

3. Results and discussion

The 2% w/w of Ag NWs were dispersed in different concentrations of IPA in water starting from 0%, 10%, 20%, 25%, 50%, 75%, 90%, and 100%. For all samples, the concentration of nanowires was kept constant at 2%w/w. Their optical images were recorded to compare the network of Ag NWs on the glass slide.

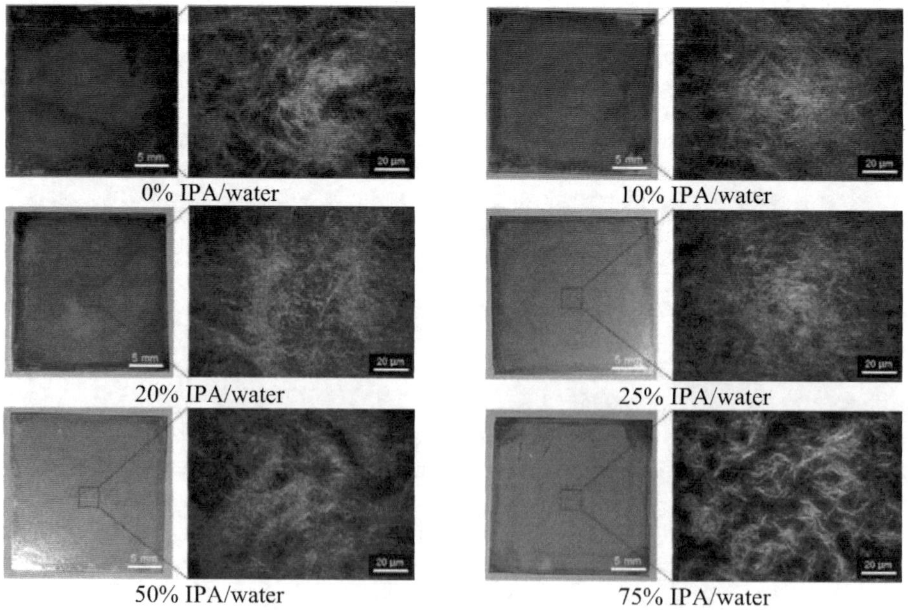

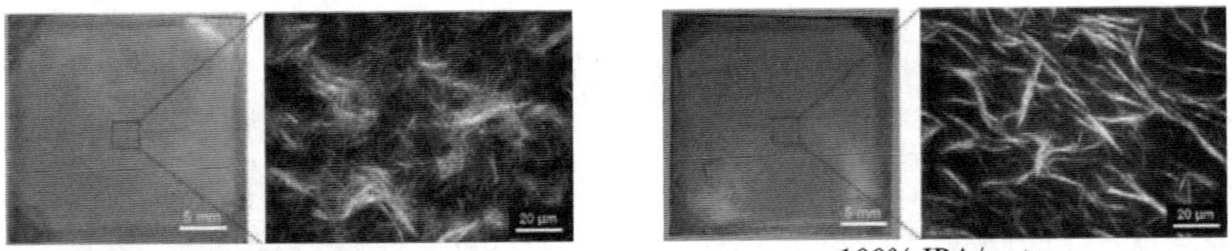

Figure 1. Optical images of Ag NWs dispersed in various concentrations of IPA prepared on PEDOT: PSS-coated glass slides.

The optical images of those IPA concentrations are shown in Figure 1. It shows Ag NWs on the glass slide by appearing in the opaque film. When increasing the magnification of the images, it appeared that Ag NWs distributed on the glass slides. The heaters can be separated into two groups, 0% to 50% and 75% to 100%. In the first group, 0% to 50%, the Ag NWs film dispersed on the surface and was connected. In contrast, at higher concentrations of IPA (or the second group), 75% to 100%, some locations appear to have no interconnection with the Ag NWs network. The bundle of nanowires may result from the fast evaporation of IPA rather than water which make nanowires aggregate easily.

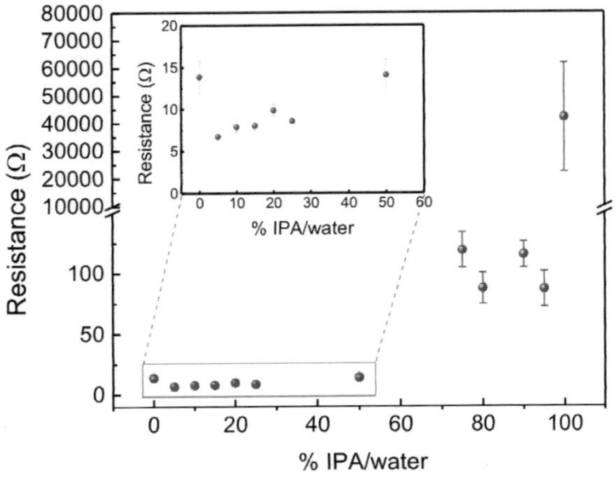

Figure 2. the electrical resistivity of the heater prepared with Ag NWs dispersed in IPA mixed with water at different ratios (% IPA/water).

All spun nanowires on the glass slide were measured in their electrical resistance, resulting in Figure 2. The results show that when the %IPA/water is less than 75%, the electrical resistivity appears to be less than 15 ohms. When the IPA/water ratio is greater than 75%, the electrical resistivity exceeds 85 ohms, and it can reach 42 kilohms when Ag NWs are dispersed in 100% IPA. This discovery demonstrated that DI water could improve Ag NW adhesion to glass. The results are consistent with the optical image in Figure 1. In addition, the presence of the PEDOT: PSS layer also contributes to the complete network formation of Ag NWs.

Based on the resistivity and dispersion of Ag NWs in Figure 1 and Figure 2, a heater prepared with Ag NWs dispersed in 25% and 50% IPA/water was chosen because the films are better dispersion of Ag NWs and smoother than the other % IPA/water. The resistivity is low enough to be a heat source at low power. The

DS1820 sensor was used to measure average temperatures on both heater constructions of PEDOT: PSS/Ag NWs at 25% and 50% IPA/water. Voltages ranging from 1 to 10 V were applied, and the results are shown in Figure 3a.

The results of thermal generation from PEDOT: PSS/Ag NWs heaters prepared from Ag NWs in 25% and 50 % IPA/water were supplied with continuous voltage in the range of 1 to 10 V for 30 min at each voltage. The average temperatures are shown in Figure 3b to demonstrate the constant temperature value of heat generation. It can be seen that the heaters from Ag NWs distributed in 25% and 50 % IPA/water can be heated almost equally over the entire voltage range. The heater's temperature increases from room temperature (25 °C) to 60 °C (at a voltage of 10 V).

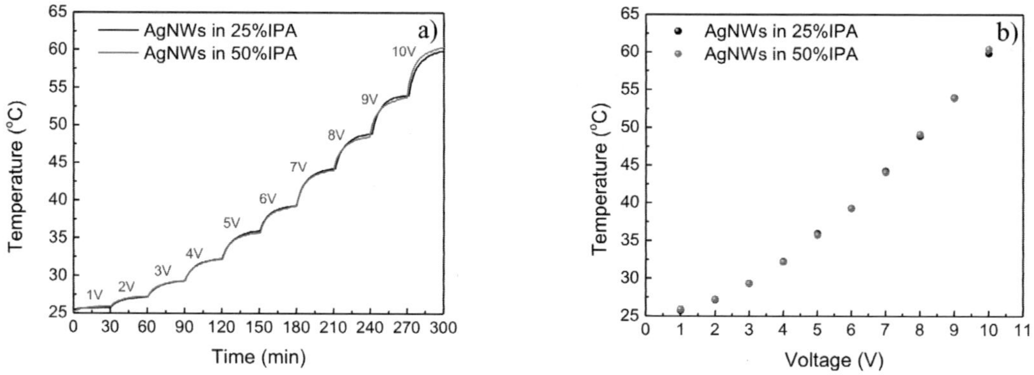

Figure 3. The average temperature of PEDOT:PSS/Ag NWs heater under the applied voltages of 1-10 V.

Next, the heaters were measured for their heat characteristics. Figure 4a shows the temperature measurement results from a heater prepared with Ag NWs dispersed in 25% and 50 % IPA/water with a constant voltage of 6V applied for 30 min to heat the heater and stop the power supply letting the heater cool down for 30 min. The highest temperatures of the heaters were 38.44 °C and 38.04 °C, where Ag NWs were dispersed in 25% and 50 % IPA/water, respectively, and could be reduced to room temperature after the power supply was stopped. From the results of Ag NWs dispersions and their electrical resistance, Ag NWs in 25% IPA/water demonstrated better overall properties than Ag NWs in 50% IPA/water.

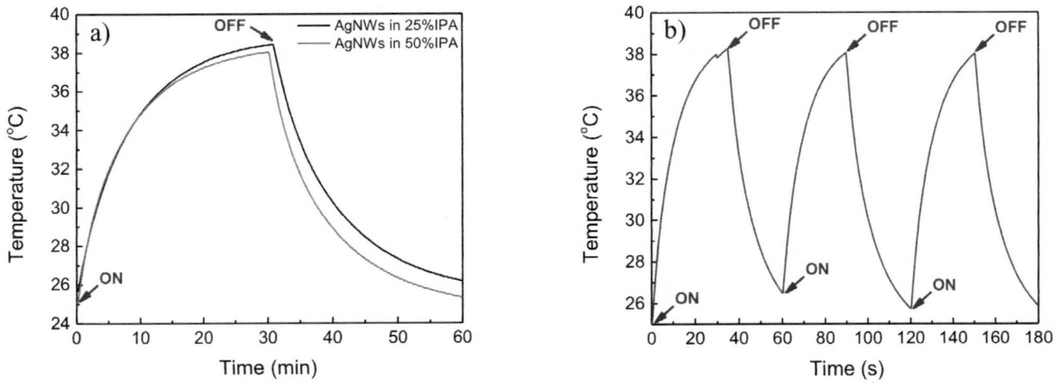

Figure 4. The temperature at different times of PEDOT:PSS/Ag NWs heater (a) and thermal repeatability with its average temperature (Ag NWs in 25% IPA) (b), both pictures are under an applied voltage of 6 V.

Therefore, the heater prepared with Ag NWs dispersed in 25 % IPA/water was then used for more investigation. The temperatures under applied voltage at 6 V for 30 min and with no voltage for 30 min were collected for three cycles, as shown in Figure. 4b. The result shows an average measured temperature was 38 °C, and the heater could be reused continuously.

4. Conclusions

The heaters were investigated by varying 2%w/w of Ag NWs in different concentrations of IPA in water. The findings indicate that DI water can improve Ag NW adhesion to glass which helps improve the low resistivity of the heaters. The heaters in the first group, 0% to 50% IPA, had resistivity less than 15 ohms when voltages of 1-10 V were applied to them. The heater of PEDOT: PSS/Ag NWs with 25% IPA shows the best results. The heater can be repeatedly used, which could be suitable for the heater application and other applications.

Acknowledgment
This work is supported by Spearhead Project through Kasetsart University and Treasure Products Co., Ltd.

References

[1] Ji S, He W, Wang K, Ran Y and Ye C 2014 *Small.* **10** 4951.
[2] He W and Ye C 2015 *Mater. Sci. Technol.* **31** 581.
[3] Hong S, Lee H, Lee J, Kwon J, Han S, Suh YD, Cho H, Shin J, Yeo J and Ko SH 2015 *Adv. Mater.* **27** 4744.
[4] Gupta R, Rao KDM, Kiruthika S and Kulkarni GU 2016 *ACS Appl. Mater. Interfaces.* **8** 12559.
[5] Papanastasiou DT, Schultheiss A, Muñoz-Rojas D, Celle C, Carella A, Simonato J-P and Bellet D 2020 *Adv. Funct. Mater.* **30** 1910225.
[6] Patel M, Seo JH, Kim S, Nguyen TT, Kumar M, Yun J and Kim J 2021 *J. Power Sources* **491** 229578.
[7] Li J, Liang J, Jian X, Hu W, Li J and Pei Q 2014 *Macromol. Mater. Eng.* **299** 1403.
[8] Pyo K-h and Kim J-W 2016 *Compos. Sci. Technol.* **133** 7.

AUTHOR INDEX